Illustrated by Susan Zises

DAVID McKAY COMPANY, Inc.
NEW YORK

The Book of Gadgets

COPYRIGHT © 1974 BY Tania Grossinger

All rights reserved, including the right to reproduce this book, or parts thereof, in any form, except for the inclusion of brief quotations in a review.

LIBRARY OF CONGRESS CATALOG CARD NUMBER: 74–82984
CLOTHBOUND ISBN 0–679–50505–9
PAPERBOUND ISBN 0–679–50506–7
MANUFACTURED IN THE UNITED STATES OF AMERICA

Designed by Bob Antler

The Book of Gadgets is dedicated to
Daniel M. O'Shea, my friend and agent,
the "man without whom...."

ACKNOWLEDGMENTS

This book could not have been written without the very real cooperation and support of many people. I would like to extend special appreciation to: Mrs. Heidi Fischer, Director of Public Relations, Hammacher Schlemmer, Inc.; Mr. Dominic Tampone, President, Hammacher Schlemmer, Inc.; Mr. Richard Hochman, Public Relations Manager, National Housewares Manufacturers Association; Mr. Leonard Silver, President, Hoffritz, Inc.; Mr. Donald Smith, Director of Press Information, Abercrombie & Fitch; June Gittelson, Director, The Left Hand; Jack Mazotas, Manager, Bazar Français; The Shelburne Museum, Vermont; Marion and Charles Klamkin; Stan Lee; Larry Penzell; Fred Pohl; Dr. Edward Stephens; Roberta Kopper; Lou Weinstein; Art D'Lugoff.

CONTENTS

1	**The Great Gadgetsby** *What, When, Where, Why—And Why Not?*	1
2	**Be It Ever So Humble** *Gadgets for the Home*	35
3	**The Gadgeting Gourmet** *Gadgets for the Kitchen*	42
4	**I Get High with a Little Help from My Friends** *Gadgets for the Bar*	52
5	**An Ounce of Prevention** *Gadgets for Your Health*	57
6	**I Get By with a Little Help from My Friends** *Gadgets for Grooming*	66
7	**How to Succeed in Business Without Really Trying** *Gadgets for the Office*	72

8	Getting There Is Half the Fun *Travel Gadgets*	82
9	A Sport Is a Sport Is a Sport *Gadgets for Fun and Games*	93
10	Mother, Please, I'd Rather Do It Myself *The Do-It-Yourself Gadgets*	101
11	2001 *Gadgets of Fantasy and the Future*	123
12	Where to Find Them *Resource Section*	139
	Index	153

The prices in this book were accurate at the time it went to press, but of course they may vary with time due to market fluctuations.

THE GREAT GADGETSBY
What, When, Where, Why— And Why Not?

What is a gadget? Ask ten different people and you'll get ten different answers. Dictionaries don't tell us much. Webster's describes it as "any interesting, but relatively useless or unnecessary object." Funk & Wagnall's adds "any small mechanical device or contrivance, especially one of which the name cannot be recalled." Neither of these definitions, to my way of thinking, does justice to the gadget.

Gadgets masquerade under many names. As the dictionaries point out, they are often referred to as devices or contrivances. Yet they are also known as gismos, contraptions, whatchamacallits, thingamajogs, even gimmicks.

They simply can't be understood by knowing what they are called. They must be studied in reference to how they function, and why. Basically, a gadget *does* something, though not necessarily a something that couldn't be done just as well without it. Yet a gadget has a very real raison d'être. It invariably saves time, saves energy, saves money, and simplifies the task to be done, thus freeing the owner to do other things.

An attempt to define gadgets often results in a contradiction of terms. Gadgets can be practical or unpractical, disposable or indisposable, useful or useless, expensive or inexpensive, simple or complicated, battery operated or manual, dispensable or indispensable, lightweight or heavy, stationary or portable, relevant or irrelevant, or a little bit of all these things. More often than not, they are amusing, fun, perhaps a bit ridiculous, imaginative, innovative, unique, ingenious, incredible, fashionable, and, often as not, "in."

For our purposes, it isn't necessary to consider whether or not a gadget ceases to be one once it has gained wide acceptance and is considered a staple item, or whether an appliance can still be considered a gadget. Forget the semantics. What really counts is that gadgets play an important role in our day to day living and, as such, deserve our attention. But if you must define it, it's really quite simple. If you consider it to be a gadget, it is.

The gadget boom in America began shortly after World War II at a time when people had a need to escape the restrictions formerly imposed on them by the war, had more time to try new things, and, in a period of a boom economy, had more money to play with. Today, gadgets are the foundation of a multimillion-dollar business, and it might be interesting to discover who buys what—and why.

New York's famous Hammacher Schlemmer finds that most of its customers are women between the ages of thirty and fifty who spend an average of twenty-five dollars per visit and come back often. Leonard Silver, president of the national chain, Hoffritz, Inc., estimates that 46 percent of his clientele is male, 34 percent female, and the remaining 20 percent couples. But he acknowledges that his stores are more male-oriented than most.

3 • The Great Gadgetsby

The young seem to be just as fascinated by gadgets as their parents, if not more so—especially by anything electronic. Their influence on the economy at large is well documented, and they seem to be more than willing to spend money experimenting with new things.

Wealth is not the important factor it may have been twenty-five years ago. Gadgets have long ceased to be looked on as "rich men's playthings." People with a lot of money may shop at Abercrombie & Fitch and other specialty stores; those who don't may join the hundreds of thousands who buy through the hundreds of mail-order catalogs available. One way or another, gadgets are available to just about everybody at a price he can afford.

While it is true that some people buy gadgets for the obvious practical reasons that they save time, energy, or money, for others, the reasons are more complex. Dominic Tampone, president of Hammacher Schlemmer, shares some of his insights:

> People in the leisure class have all their basic needs taken care of. So the question is, what can you tantalize them with next? They not only need "something" to talk about at their next party, they need the *latest* "something." In addition, gadgets satisfy the toy-desires of the individual. They may, in fact, be practical too, but basically they represent an emotional reaction to wanting something else to play with. This is more true for men, who still have the "little boy wants a toy" need. Even when a man comes in ostensibly to buy an item for his wife, what he really is buying is a toy for himself. Men never grow up. Women, on the other hand, are more practical. They want useful, pretty, lovely things, not just playthings. A woman looks at a gadget and says, "Another thing to clean." A man says, "What fun."

Leonard Silver of Hoffritz believes the status role of gadgets motivates a great many customers. "Luxury and unique household items are big nowadays, because in an era when everybody can and does travel to Europe and the Far East as if it were around the corner, the home is about the only place left where a person can show off his status symbols."

In certain situations, however, gadgets may become more than mere fun or a convenience. An electric can opener, for example, may be "just another gadget" for many, but to a person suffering from arthritis, it is a godsend. The success of The Left Hand store in New York City is based on a very real need: for the one out of every five Americans who is born left-handed, gadgets like left-handed corkscrews and can openers or pie and cake servers with serrations on the left make it easier to cope with what is otherwise a counter-clockwise world.

Then there are the "gadget freaks," who carry their interest to an extreme. Some collect for the sheer joy of collecting. To others, collecting has become almost an obsession: they must have every new item on the market relevant to their special area of interest. One gadget freak recently told me, only half-humorously, that she regarded scientists as men who prolong life just to make sure she'll have time to pay for all the gadgets they invent.

The gadgetmongers differ when asked if their customers actually use all the gadgets they buy. Leonard Silver refers to his novelty gadgets as FADs, meaning they are used For A Day. But in the case of the more popular run-of-the-mill items, he believes that most people who buy them really do use them. "It's part of the new life," he says. "In this age of specialization, we encourage special items for special services. One no longer cuts paper with sewing scissors even though, in fact, one could. Instead, one buys special paper scissors to do the job."

Jack Mazotas, manager of the Bazar Français in New York City says, "People use them as long as the fascination lasts. When it stops, they usually end up in a drawer somewhere."

5 • The Great Gadgetsby

Dominic Tampone adds, "Basically, if you are a good cook, all you really need is three pans. Most of the items we sell end up on the top shelf of the closet a month later."

Yet when one looks at the sales figures for some of the best-selling items, it is hard to believe that so much money is being spent on products that aren't being used. An outstanding seller at the Hoffritz stores throughout the country is a simple mustache comb that clearly owes its success to its adaptation to a specific trend in present-day society. Originally developed as an eyebrow comb, the sales were less than spectacular until mustaches became "in." The Eyebrow Comb was renamed the Mustache Comb, just as the Eyebrow Razor was remarketed as a mustache razor. Both became staggering successes.

The Swiss Army Knife has been a perennial best seller over the past five years because, though physically only one item, it has fourteen separate functions. Barely three and one-half inches long, it contains a screwdriver, hole punch, Phillips head screwdriver, nail file, cap lifter, tweezers, toothpick, metal saw and file, fish scaler and disgorger, saw pen blade, scissors, spear blade, shackle, and can opener. No other single gadget combines so many features.

In the early 1960s, the most popular kitchen gadgets were the very simple ones; vegetable steamers, lemon strippers, butter curlers, and onion and tomato slicers. Today they are more complex. Dominic Tampone predicts Hammacher Schlemmer's next best seller will be the self-stirring saucepan, because, in principle, it frees the cook to do other things during the time he or she would previously have had to spend stirring the pot. This special saucepan also solves the problem of "how?" that arises when a recipe states, "Keep covered at all times, stirring occasionally." He suggests that the next variation will be glass covers with automatic stirrers that will not only fit over regulation-size pots, but will let the cook see what is happening inside the pot at any given time.

Two other big money grossers are the Automatic Plant Quencher (over 80,000 were sold last year at Hammacher

Schlemmer alone) and the Damp Chaser. The Plant Quencher releases the exact amount of water a plant needs in relation to the amount of heat and humidity present in the room at any particular moment. The Damp Chaser, a boon for those who have summer homes, is a long rod with a 60-watt bulb at the end; the device produces enough heat to dry out any mildewed or musty closet.

It's difficult to tell in advance just which gadgets will catch on, and which won't. However, all salesmen agree that as in every other business, publicity and promotion play a very important role in determining the success or future of a new gadget.

The Dione Lucas cooking show, launched on television in 1950, had a profound effect on the kitchen-gadget market. When Lucas used a certain kind of mincer on Tuesday, stores knew they would sell out the item by Wednesday. She exposed viewers to gadgets that prior to the existence of the show they may not have known existed. Even today, if Graham Kerr (the "Galloping Gourmet") or Julia Child uses a gadget on TV, the success of the product is ensured.

The press has an influence as well. Jack Mazotas tells of a vegetable slicer/cutter called the Mandolin that was on display recently at the Bazar Français. Because it was so cumbersome and complicated, he assumed it would be of interest only to the professional chef. *Gourmet* magazine ran an article referring to it, and since then the store has not been able to keep it in stock. Magazines, newspaper cooking columns, and cookbooks also play an important role in making the public aware of what tools are available.

Before taking a closer look at the gadgets we surround ourselves with today and what we might expect in the future, perhaps it is appropriate at this point to look back into history and discover just how gadgets came to be and what they were like in the past.

The actual word *gadget* derives from the French word *gachette*, meaning a piece of machinery. The term was used in England long before it became popular here, and gadgets per se, of course, have existed in one form or another since the beginning of recorded time.

One of the first gadgets very likely was the simple wooden skewer, created out of man's need to cook food over an open fire without burning his hands. It had other advantages too in that it could be stripped from a tree, was disposable, and the price was right.

Stone tools used for hacking go back over 100,000 years to the Early Stone Age, when large tracts of the Northern Hemisphere were still covered by ice. Most of these were simple, crude stone flakes that were used as scrapers or knives, but unlike the quickly discarded implements of the ape, man kept them for reuse. Eventually, they were used for more than one function, as happens with most gadgets. They provided a way for man to get meat off an animal without using his hands, a major step toward civilization; they functioned as simple weapons; and they were used as implements for artwork, which is how we know today that they existed in the first place.

The first axe, a pear-shaped stone flaked to a cutting edge at its smaller end and held cupped in the palm, was the major tool of the next seventy-five thousand years, with improvements like a handle and different-sized blades added on over the years.

The Neolithic period (6,000 B.C.–4,000 B.C.) was when gadgets really began to proliferate. In order to secure food with the least amount of effort, men devised traps. To clothe their bodies, they created the loom, on which they could weave fabrics out of animal skins. The first potter's wheel was discovered around this time in Mesopotamia. It was in this era, too, that the first instrument for killing at a distance, the bow and arrow, came into creation.

The wheel was invented in Southwest Asia during the early Bronze Age, though records show wheeled toys earlier than that in both Mesopotamia and ancient Mexico. The first cart with wheels was found in Syria and Sumeria around 3500 B.C., and Egypt boasted the first spoked wheel in 1800 B.C.

The forerunner of one of today's most popular gadgets, the calculator, was the abacus, a frame supporting a number of parallel rods on which colored beads were strung. It was the first mechanical method of counting. Though most people attribute the abacus to the Chinese—and they did have them before the Christian Era—Herodotus recorded its use in ancient Egypt even before that, and it was also used by the ancient Greeks and Romans.

Some gadgets that we might think of as being only a few hundred years old at the very most are really much older. Mortars and pestles for grinding household grains and spices were in use in Mesopotamia and Old Babylon as far back as 2500 B.C. The whistling jar, ancient ancestor of the whistling tea kettle, was known to have existed in 200 B.C. as part of the Mochica culture in Peru. And the Musée des antiquités nationales in Saint Germain, France, shows bronze pastry cutters ranging as far back as 200 A.D.

Gadgets in the medieval period, especially the thirteenth-century feudal epoch in France, took on a more sinister look. At that time, they were used primarily as instruments of torture. It was during this period that the guillotine, the rack, thumbscrews, and special two-headed swords gained notoriety. When gadgets weren't being used to destroy, they were used to protect. Jousting "games" were the chief recreation, and as a result, devices like the nasal, a metal bar to protect the nose during tournaments, and various helmet attachments to ensure that the winner wouldn't be wounded too badly in the head were developed.

The man many consider to be the godfather of gadgets was a man of many interests. Best known as a painter and sculptor, he was also an architect, musician, art critic, civil-military engineer, botanist, astronomer, geologist, and student of anatomy. Leonardo da Vinci epitomized the fifteenth-century Italian Renaissance spirit of intellectual curiosity and vitality. Notwithstanding his talents as an artist, as an inventor alone, we owe him a great deal. He was the first to imagine the possibility of a flying machine and parachute. His ideas for an aerial screw led to the subsequent development of the propellor and helicopter. While serving as military engineer to Cesare Borgia, he was lauded for creating a device that would enable soldiers to scale fortress walls. In weaponry, he discovered that metal projectiles could be aimed more accurately than heavy stone balls and, as a result, developed the concept of the first machine gun. Perhaps his most far-reaching discovery was made in 1482, when he first described the camera obscura, the first pinhole camera in which a tiny hole in the wall of a darkened chamber took the place of a lens.

In America, two of the most prominent gadgeteers were the third president of the United States, Thomas Jefferson, and Benjamin Franklin, signers of the Declaration of Independence.

It was Jefferson who, in the late 1700s, designed the first wall clock that showed the day of the week as well as the hour. He then went one step further; he wanted to find a way, other than by standing on a chair, to get to the clock in order to wind it once a week. So in a sense, one gadget begat a second gadget. He finally came up with the first folding ladder, in this case fourteen feet long and similar in principle to the pole ladder developed earlier by the British in India to ease climbing on and off an elephant's back.

A lover of fine wines, he wanting something to ease the task of physically carrying the bottles from one part of his house to another. This resulted in the first dumbwaiter, which could mechanically carry the wines from the cellar to his dining room whenever he pressed the appropriate button.

Because he was a music lover, he would often invite chamber-music groups to play for his guests at Monticello. Finding it cumbersome to set up five separate stands, he devised the first portable music stand that could serve a quintet. Four adjustable slanted music rests would unfold from a box for each of the instrumentalists standing on each side. The music for the fifth musician was placed on the stand centered on the top of the box, and when the concert was over, it took but a minute to put everything back in order.

An avid reader, Jefferson was perplexed with the problem of how to store and display the six thousand books he kept in his library. He finally came up with the idea of "stacking book shelves," with five-sided rectangular "book boxes," a concept regarded as modern even today.

Benjamin Franklin, a school dropout at the age of ten, is best known in his role as an inventor for having created the first printing press in 1722 and the lightning rod thirty years later, which proved that lightning is, in fact, electricity. He also gave his name to a special kind of stove, the first to give out more heat and at the same time use less fuel. Another of his well-known contributions was the bifocal lens, which permitted people to both see at a distance and read without having to change glasses. Lastly, he is credited with bringing the first bathtub into the country, from France. Shaped like a bog shoe, it had a water-heating device built into the heel and a spigot for draining water in the toe. When not in use, it was stored in the cellar.

It is interesting to note from a historical perspective that as a founding father of a country dedicated to the principle of free enterprise, Franklin never sought to protect his own inventions with patents, though the principle of patenting had been part of the legal system as far back as 1646. Rather, he believed, "As we enjoy great advantage from the inventions of others, we should be glad of an opportunity to serve others by any invention of ours, and this we should do freely and generously." But being a realist and understanding that not everybody would share his point of view, he did his part to ensure that the U.S. Constitution would legally safeguard the rights of inventors. "Congress shall have power to promote the progress of science and useful arts by securing for limited time to authors and inventors the exclusive right to their respective writings and discoveries" (Article 1, Section 8). That period of time, as it now applies to inventions and gadgets, is seventeen years.

At the same time that Benjamin Franklin was helping to write the Declaration of Independence, the United Society of

Believers in Christ's Second Coming, better known as the Shakers, were settling near Albany, New York. They were a rather large communal religious sect who produced goods not for profit but for their own use or for sale so that they could purchase more land throughout the colonies as their membership grew. The first group to create and use gadgets on a full scale, they are credited with the first cheese press, circular saw, pea sheller, automatic apple parer, fine sieve for sifting flour, the common clothespin, and, later on, the more popular wiresnap version.

One of their biggest fears was destruction by fire, and accordingly, they used their ingenuity to come up with lanterns made with covered glass panels, instead of the open-sided ones used by others during that period, and lamps with two funnellike shades and chimneys, so smoke fumes would be vented outdoors or into the stove chimney rather than indoors. They also saw to it that ceiling fixtures with candles had scalloped trays underneath to catch any hot wax that might drip down. Naturally, as they moved into new developments, they took their new gadgets with them, spreading the gospel, so to speak, in more ways than one.

One of the problems with making new products in the colonial era was that factories were unknown and distribution, in the marketing sense, unheard of. There were some Yankee peddlers, mainly Germans, Poles, Jews, and Armenians who had immigrated in the seventeenth century and made a living selling needles, pins, buttons, spoons, and knives smuggled in by English sailors, but they were few and far between. The one solution, both inside and outside the Shaker community, was to turn to the local village blacksmiths. Everything they did was made to order, and it is to them that we owe the first iron strainer, the first trivets to keep hot pots from scorching table surfaces, various types of cutlery, and the first makeshift toaster, which made use of a rotating rack to simplify the process of turning the bread over the fire.

During this period, the only place that could vie for

attention with the smithy was the general store, a place that doubled as a small commercial outlet and a center for social gathering. The managers had to create their own gadgets to fill the special needs of running a general store effectively, and things like string dispensers, twine holders, and wrapping-paper dispensers soon made an appearance. If plug-tobacco cutters, used to cut a "chaw of tobacco" from the barrel, were one of the more popular items, the cigar cutters were among the most interesting. They were not only practical for trimming the ends of cigars but, as are so many gadgets, they were fun, and their owners considered them status symbols. One, a King Alfred Cutter was thirteen inches tall and contained a Waterbury clock. Another had a counting device that told how many times a day it was used. Still later, a mechanical type operated by a mainspring and blade had a hole in the top of the dome with a warning that read, "Be Careful Of Fingers."

Much of the activity in America of the 1800s was taking place on the ever expanding frontier, and quite a few gadgets reflect this. One of the most practical ones for cattlemen were the lightweight brand rings, copper hoops that were able to run any kind of brand and be manipulated with a stick handle. Their great advantage over the heavy branding irons previously used was that they could be conveniently tied to a saddle string and were easy to transport on horseback. Another cattleman favorite was the barbed-wire stretcher.

One of the most important gadgets in frontier America that really signified a breakthrough in terms of land expansion was the miner's candlestick, developed in response to the problem miners had when they needed light to see in order to dig into the sides of mountains and deep into the mines. It took some time before a practical solution was found, but found it was in the form of a candle that was inserted into a cupped receptacle with a spike and then stuck into the timber, shoring posts, or rock crevices.

As the country continued to expand geographically toward the end of the nineteenth century, certain changes were taking place in the living patterns of the people. Perhaps the

13 • The Great Gadgetsby

greatest single offshoot of the rise in the standard of living at the turn of the century was the advent of the mail-order catalog. The first one, Montgomery Ward's Catalog of 1883, pledged to provide something for everybody. "We are prepared to help you walk, ride, dance, sleep, eat, go to church on time, or stay at home." Suddenly, there was a mass audience and a way to reach them, through advertising and marketing. And it is at this time that gadgets really came into their own as a way of life for both the consumer and the manufacturer. In 1895, the catalog was over six hundred pages and listed over fifteen thousand items in thirty-eight categories including dry goods, books, artist's supplies, photographic and printing equipment, harnesses, wallpaper, saddlery, carpets and curtains, tinware and cutlery, dairy supplies, bicycles, baby carriages, woodenware, agricultural implements, carriage hardware, and musical supplies. Here are a few that could be ordered that year:

A woman wanting a hair brush could buy the plain familiar one or order a more exotic model ("the most powerful magnetic brush ever made; cures headache, neuralgia, rheumatism, prevents hair from falling out, and aids in restoring gray hair to its natural color. In the back of this brush are embedded strongly charged magnets in contact with which are the fine, highly polished steel wires which form the brush so that when applied to the head or flesh, currents of magnetic electricity flow from the magnet through the wires"). This brush sold for $1.25, plus an extra 25¢ for "the new improved model with an extra magnet in the handle."

Housewives could find meat choppers, coffee grinders, sausage stuffers, fruit and lard presses, presses that made fruit, wine, and jelly, wringers for the washday, throat and nasal atomizers, harmonica holders, or gasoline stoves. At a time when people made their own milk and few had their own refrigerators, there were milk coolers and aerators "especially designed to cool and aerate new milk or cream from separator, with or without the use of ice. . . . at the same time removes animal and garlic orders."

No one who raised sheep could afford to be without the

Hero Sheep Protector—at least that's what the ad would have us believe. It "is made of steel galvanized wire formed into links. Each link has two sharp projections. Each collar consists of thirteen links which will reach around the ordinary sheep's neck. . . . sheep cannot hurt themselves with these protectors. By the use of this protector, 95% of the sheep killed by dogs, wolves, etc. would be saved. You say that a dog does not always catch a sheep by the neck. We say, right you are; but when they catch them elsewhere it is done to check the sheep so that they can get to its neck. . . . in some instances the object in catching the sheep is to cut the throat and drink the blood. In their wrestle with the sheep, they are sure to come in contact with this protector. This closes the chase; they will not give blood for blood. If you will put the Hero Protector on your sheep, you can pasture them in your remotest field, and you need not lay awake at night for fear of them being molested."

A look at the competitive Sears Roebuck catalog of 1897 supports the fact that vanity knows no century. The Princess Bust Developer at $1.46 was designed to "build up and fill out shrunken and undeveloped tissue and form a rounded, plump, perfectly developed bust, producing a beautiful figure", and Madame Schack's Dress Reform Abdominal and Hose Supporter, "recommended to reduce corpulency and to all who suffer weakness from their sex," were two of the hottest items.

Sears Roebuck, as a rule, did not experiment with fads. They claimed that the items they displayed represented items people really wanted, bought, and used. Some of them include ear trumpets and conversation tubes, Health Inhalers ("an improved breathing tube with valve for systematic chest expansion by use of common air and the cure of diseases of the lungs, throat and other air passages, and disorders of circulation, digestion and assimilation"), simple padlocks ("a lock is something you buy to wear out"), Heaton's patent button machine ("any child can now fasten the buttons on his own shoes"), combination umbrellas and canes, and multi-objects like the Eagle combination pocket pencil, which included a compass, pencil-point protector, pencil holder, envelope opener, and rubber eraser, all for eight cents.

In the eighteenth and nineteenth centuries, few people had the creature comforts we accept today as necessities of life. Though we might find it difficult to imagine living without them, many people did so, and when many things we now take for granted first came on the scene, they were very much regarded as gadgets and, more often than not, gadgets only for the well-to-do.

The flush toilet is a case in point. Actually, the first flush toilet with a wooden seat has been traced to the royal palace at Knossos, Crete, circa 1800 B.C. Then it disappeared for centuries. Though the first patent for it was issued in London in 1775, even in the 1850s most people there were still using the chamber pot, which had to be emptied out doors. It wasn't until the 1870s that the toilet finally began to gain acceptance in America.

Similar is the case of bathtubs and showers. Bathtubs, though they existed during the American Revolution, didn't enjoy a room of their own until after the turn of the century and even then only by very few. Most people used what was known as a Virginian stool shower, which required a pail or basin as its source of water, to wash themselves. Made of walnut, with a revolving seat resembling a piano stool, a lever on one side was worked back and forth to pump the water up through a hose attached at the back of the shower over the bather's head. Simultaneously, hand action worked a scrub brush up and down the bather's back. One had to work up quite a sweat in order to keep clean.

The most practical bath, for those who took it in the kitchen, were the closet tubs, which could be opened and shut at will, much like the famous Murphy bed. Finally, in 1888, tubs were developed that had their own built-in needle spray and shower. The Glamor Tub boasted ten vertical cage bars that could squirt water at the bather. It actually wasn't until after World War I that showers became permanent fixtures and the bathroom-fixture business itself began to boom.

Irons, or gadgets used to perform the same function, go back as far as the tenth century, when the Vikings used glass objects that looked like flat, inverted mushrooms to smooth their linens. The Greeks had a gaffering iron to pleat their linens and robes, and other people in other places in other periods relied on flat mallets or paddles called hand mangles to beat clothing free from wrinkles. In sixteenth-century Europe, ironing was done with a hollow, boxlike object that was designed to hold hot charcoal and evolved a century later into an iron designed to hold a separate slug of metal, which was inserted after being heated in an open fire.

The year 1895 marked the appearance of gas as well as acetylene, alcohol, and gasoline irons, the latter two coming with attached tanks to hold the alcohol or gas. Though the first electric iron was patented in 1882, it didn't come into its own until 1927, when the adjustable thermostat became a permanent fixture on it. A year later, the first steam iron was introduced, but it wasn't until ten years later that one was made that could be set on its heel without spilling water. Improvement was added to improvement until today one has a myriad of choices of self-cleaning steam, spray, and dry irons available within everybody's budget.

In 1811 in England, Jane Hume took out a patent for the first machine that was used for sweeping floors. It consisted of a box equipped with a brush turned by a pulley and string mounted on the box's broomstick handle and was more trouble than it was worth. There were high hopes for a later model, but unfortunately, when put in use, it stirred up more dirt than it managed to collect, so people stuck to cleaning their carpets by hand with whisk brooms. At last in 1876, Mel Bissell designed a carpet sweeper that actually worked—on floors as well as carpets. This proved to be the forerunner of models still in favor today.

The first vacuum cleaner, introduced in 1901, was large, clumsy, and cumbersome. It was finally improved a decade later with the addition of a powerful but lightweight motor.

17 • The Great Gadgetsby

The first portable arrived from Sweden in 1924 with a flexible hose that could attach to the nozzle. This made it possible for the first time to clean furniture and draperies, something that couldn't be done with the upright cleaner. Other attachments introduced later included nozzles that loosened the dust via vibrations of pulsating air, shag rakes designed especially for deep-pile rugs, self-sealing dust bags, and devices that turned the vacuum off automatically as soon as the dust bag was filled. Today, the lightweight portable cleaners available do just about everything but turn themselves on.

The most popular gadgets in terms of volume of business and variety of selection, a century ago as well as today, seem to be those associated with the kitchen. As mentioned before, appliances that we consider staple items in our home such as can openers, pressure cookers, broilers, waffle irons, toasters, mixers, and blenders were not always thus. In many cases, interestingly enough, it took quite a number of years before they were even welcomed into the household, much less fully integrated into day-to-day living.

One of the kitchen gadgets one would have thought had always been popular is the simple can opener. Yet history shows that even though quite a few models were available over a long period of time, it took many, many years before the public finally began to appreciate them. The patent for the first can opener was taken out in 1858, forty-eight years after Peter Duran of England took out his patent on the first tin can. During those interim years, rumor had it that the French weren't designing their bayonets as weapons at all, but, much more practically, as devices with which to open cans. The can opener of 1858 actually did look much like a combination bayonet and sickle. Twenty years later, Lyman introduced his can opener, which substituted a cutting wheel for the spike or knife blade, the basic principle for all subsequent openers. Even though it simplified the task of opening a can, few people were interested in spending their money on it. Finally, in 1930 a can opener that could be attached to a wall was introduced, and sales took off. Nineteen fifty-six saw the first can opener in combination with something else, in this case, a knife sharpener. It fell flat on its face the first three years. Not until the development of the first electric can opener in 1957 did the public's interest pique to the degree that now they are made in combination with pencil sharpeners, food grinders, scissor sharpeners, ice crushers, food mixers, juicers, salad makers, hand blenders, beaters, and mashers, and, as sales items, they can't be beat.

The first toasters, in the form of wooden tongs and skewers, were devised by the ancient Egyptians, who discovered that by scorching bread to remove moisture, it would last longer. Toasters didn't change much throughout the ages, and in the seventeenth and eighteenth centuries Americans followed the same principle and toasted their bread with the forks and wrought iron devices created by their local smithies.

The first electric toaster was introduced in 1910, and immediately there were various toaster combinations, the most popular being the Perc-O-Toaster, which served as a coffee maker as well. Another, trioed with a grill and hot plate, never got off the ground. Basic improvements were attempted in the following decade, but for inexplicable reasons they just didn't take off. In 1924, D. A. Rogers of Minnesota developed a toaster he claimed would shut off automatically, the hinged sides dropping down as soon as the toast was finished so it could be removed easily, and the owner for the first time could dial the degree of doneness he preferred. Unfortunately for Rogers, it was only a claim. In reality, the machine didn't work. A year later Toastmaster introduced their first model, but it too was plagued with problems, in this case the starting temperature of the toast, the moisture-content control, and the fact that there was difficulty reconciling the temperature with the timing.

The first "drop-down" toaster that actually did work finally appeared in 1930. From then on, toasters could only make money. Today there are vertical models that can toast different slices of bread of different thicknesses to different

19 • The Great Gadgetsby

degrees of doneness at the same time, drawer-type versions that are also used to defrost or keep things warm, and the most popular of all—toaster-baker-broiler ovens. The statistics speak for themselves. In 1922, 400,000 toasters were sold. Fifty years later, the figure went up by 7 million.

Waffle irons were far more popular hundreds of years ago than they are today. They actually date back to the fourteenth century, when in essence they consisted of two long handles so the owner wouldn't have to get too near to the heat and two heads with oblong cutouts of different sizes that shut together like pincers. This prototype survived for five hundred years, until the first wrought iron waffler was manufactured in 1860, followed by the electric waffle iron after World War I and the automatic one in 1930. There have been very few changes since then.

The first pressure cooker has been credited to Denis Papin, a Frenchman who in 1682 developed a machine he called the Digester. Experiments with it caused many accidents, and it was forgotten until 1811, when Napoleon was looking for a way to supply proper food for his fast-moving troops. One of his men, Nicholas Appert, came up with the idea of canning, thus reviving interest in the pressure method as a means of food preservation. As a result, various pressure-cooker-canners were developed and subsequently brought to America, but they didn't explode into a big business until 1917 when the U.S. Department of Agriculture announced that the only safe way to process low-acid foods was to use a pressure cooker. Improvements were made to render them less bulky and easier to manipulate. The standard electric model introduced in 1954 is pretty much the same as those sold today.

It is hard to understand, in an age where everything electric is almost worshiped (or anyway, was—until the energy crisis), why it took so long for the electric frypan to gain acceptance. First introduced by Westinghouse in 1911, it doubled in function: when turned upside down on its separate cast iron stand, it could also be used as a hot plate. Yet it didn't attract the public until 1953 when a mechanism to control heat was added. Finally, a year later, with the addition of a detachable probe that simplified the cleaning of it (the pan could now be totally immersed in water), it became a commercial success.

Nancy Johnson invented the first ice-cream freezer in 1864. The sale of these freezers reached a peak after World War I and then died down until early 1970 when, because of the nostalgia craze, they became very popular items. Interestingly enough, the hand-turned old-fashioned wooden-tub types are much more popular today than the many electrically operated models available.

Mixing, or beating, as it was known before 1907, was done by hand with whisks until L. H. Hamilton and Chester A. Beach perfected a high-speed, lightweight "universal motor" that could operate on AC or DC current. This led to the creation of the Hamilton-Beach electric mixer. A base-mounted and/or portable model was introduced in 1923, but the one that really stimulated the public's imagination was developed by Sunbeam in 1930. In addition to being a mixer, it boasted attachments that extracted and strained fruit juice, chopped meat, ground food, sliced, shredded, grated, peeled potatoes, sharpened knives, polished, buffed, opened cans, mixed drinks, and ground coffee. Today, in these and other combinations, there are over fifty different mixers to choose from.

Contrary to public opinion, the blender was not invented by Fred Waring, but by Stephen J. Poplawski, who took out his patent in 1923. He was definitely not a businessman, and if the truth be known, he only invented it to make it easier for him to make malted milks for himself. It didn't occur to him that he could use it to blend fruits and vegetables as well, and when it did, he just couldn't find a market for it. In 1938 the Waring people incorporated and came up with a somewhat similar blender that they promoted as a daiquiri-mixer bar gadget. World War II brought production to a halt, and after the war, sales came to a standstill, because it was difficult to convince people to spend so much money on something they thought of as having only one function—to mix drinks. Things picked up considerably when they began adding attachments—an ice crusher in 1956, coffee grinder in 1957, timing control in 1964, and finally in 1965 the familiar buttons that control the various speed features and other functions. Out of this came a whole new approach to cooking, "Spin-Cookery," one of its spin-offs being a whole new market for a different kind of cookbook, the blender cookbook.

When asked what the perennial best-selling kitchen gadget is, an assistant buyer for housewares at Gimbels department store replied, "The coffee maker. Everyone seems to be on a quest for the perfect cup of coffee and will spend whatever is necessary for the perfect coffee pot in which to make it. My estimate is that the average household has at least three different coffee makers on hand at any one time, and they won't hesitate to buy a fourth whenever a new model comes on the market."

Coffee was served as early as 1670 and became very popular during the Revolution, though prior to 1868, it was only available in the "green" state, that is, raw, which was far from ready for the coffee pot. First it had to be roasted in the kitchen stove and, more often than not, it ended up tasting badly at best, scorched at worst. In 1868, John and Charles Arbuckle in Pittsburgh patented a process for coating the roasted beans with a sugar glaze that sealed the pores, thereby preserving the flavor and aroma of the coffee. In 1873, Arbuckle's Ariosa coffee in the one-pound package took the country by storm.

Though the patent for a coffee mill wasn't applied for until 1892, as many as 300,000 mills were being sold yearly as far back as 1831. By 1900, specialized coffee-and-tea stores and grocery stores had their own grinders, so most people stopped buying them. With the advent of canned and vacuum-packed containers of coffee, the sale of coffee mills dropped drastically, until the past five years, when people seem to be taking a renewed interest in home-ground coffee.

The first percolator-type of coffee pot was called the biggin and was invented in France in 1800. It was a drip pot designed so that water percolated down through the ground coffee in a special compartment above the coffee pot. This was a monumental improvement over the way it was prepared before—boiling the ground coffee in water "until it smelled good."

The first percolator as we know it today, one that brews coffee below the boiling point, made its appearance in 1890, followed by the first electric percolator in 1908. One that automatically shut off after a predetermined time and temperature was reached appeared in 1931, but it had a problem keeping the coffee warm; this was later corrected.

In 1941, Dr. Peter Schlumbohm invented the Chemex coffee maker and took it to the housewares buyer at Macy's, who turned it down, telling him, "It's not a coffee maker because it doesn't look like a coffee maker." Fortunately for Macy's, they don't make that kind of mistake often. The Chemex is still one of the most popular models available today.

Soon, the market was flooded with drip coffee makers, and glass coffee makers, all best sellers. Today, electric percolators not only make coffee, they can be preset to start warming up at a specific time in the morning so the coffee is ready when you wake up. And who knows, perhaps someday someone will come up with one that will drink the coffee for you—so you can get a little extra sleep.

Finally, one can't write about coffee makers without referring to one of the more unique variations, the electric cappuccino-espresso maker made for the home. It heats and reheats the Italian coffee automatically and includes a built-in steam jet, just like the professional models, that puffs up the milk for the cappuccino.

An entire book could be devoted merely to the development of gadgets. Perhaps now it is time to take a look at what they have evolved into and the role they play in so many facets of our lives.

1. Shaker Apple Peeler, early 1800s
   —*Hancock Museum. Photo by Charles Klamkin*
2. Shaker Butter Press for forming butter into blocks, 1800
   —*Old Chatham Museum. Photo by Charles Klamkin*
3. Shaker Hand Press for squeezing fruit, early 1800s
   —*Old Chatham Museum. Photo by Charles Klamkin*
4. Adjustable, portable Shaker reading stand, early 1800s
   —*Old Chatham Museum. Photo by Charles Klamkin*

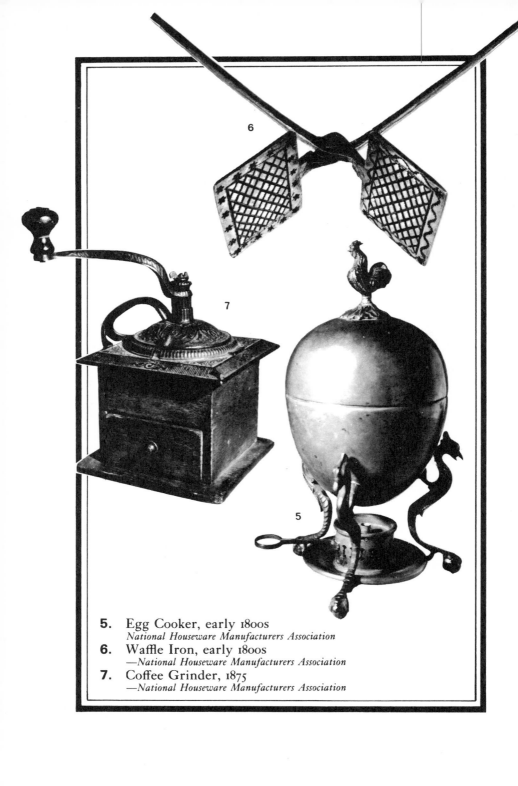

5. Egg Cooker, early 1800s
National Houseware Manufacturers Association
6. Waffle Iron, early 1800s
—National Houseware Manufacturers Association
7. Coffee Grinder, 1875
—National Houseware Manufacturers Association

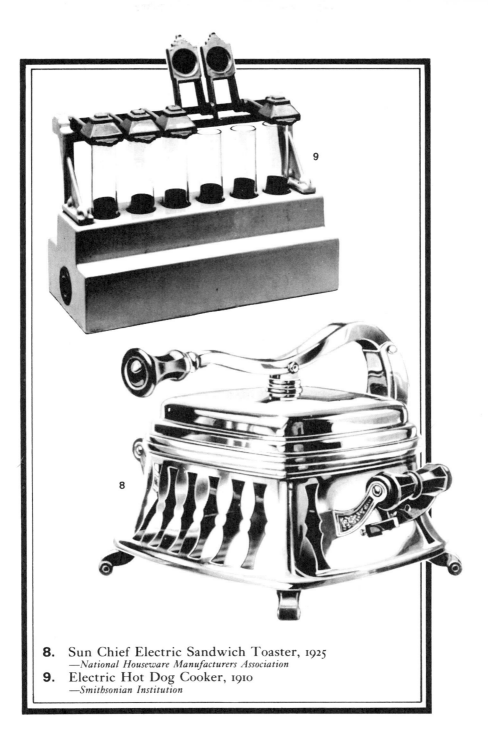

8. Sun Chief Electric Sandwich Toaster, 1925
 —*National Houseware Manufacturers Association*
9. Electric Hot Dog Cooker, 1910
 —*Smithsonian Institution*

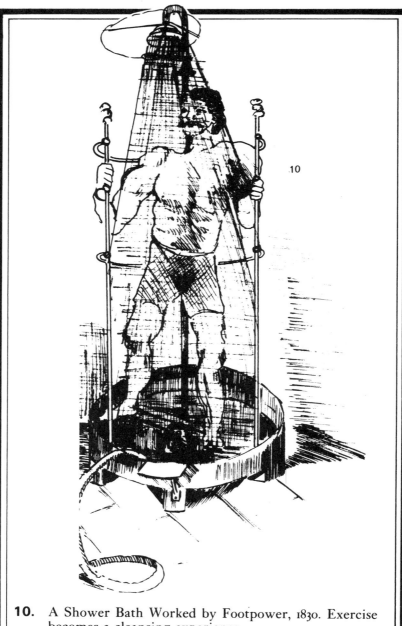

10. A Shower Bath Worked by Footpower, 1830. Exercise becomes a cleansing experience.
—*Bettmann Archive*

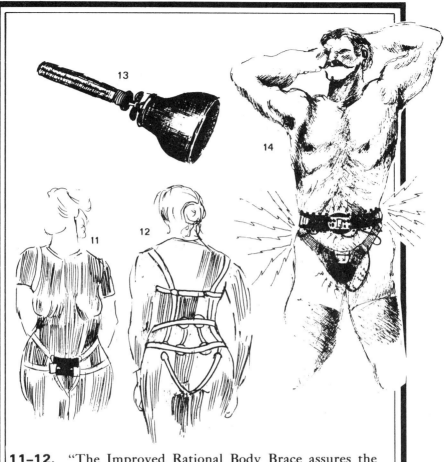

11–12. "The Improved Rational Body Brace assures the wearer comfort, vigor, health, elasticity, ease. All of the utmost importance to women subject to physical changes, in all conditions and in every walk of life."
—*Sears Roebuck Catalogue*, 1902

13. "The Princess Bust Developer sold under positive guarantee to enlarge any lady's bust from 3 to 5 inches."
—*Sears Roebuck Catalogue*, 1902

14. "Eighteen Dollar Giant Power Heidelberg Electric Belt —the suspensory encircles the organ, carries the vitalizing, soothing current direct to the delicate nerves and fibers, strengthens and enlarges the part in a most wonderful manner."
—*Sears Roebuck Catalogue*, 1902

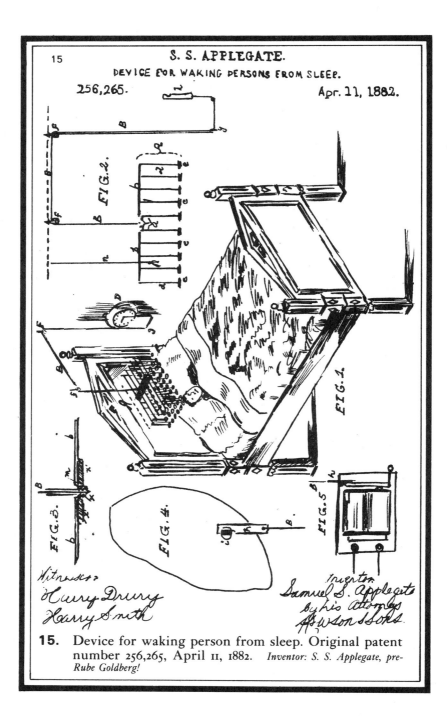

15. Device for waking person from sleep. Original patent number 256,265, April 11, 1882. *Inventor: S. S. Applegate, pre-Rube Goldberg!*

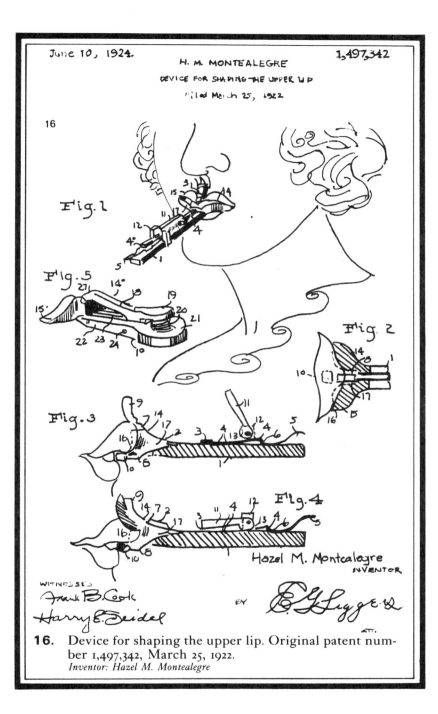

16. Device for shaping the upper lip. Original patent number 1,497,342, March 25, 1922.
Inventor: Hazel M. Montealegre

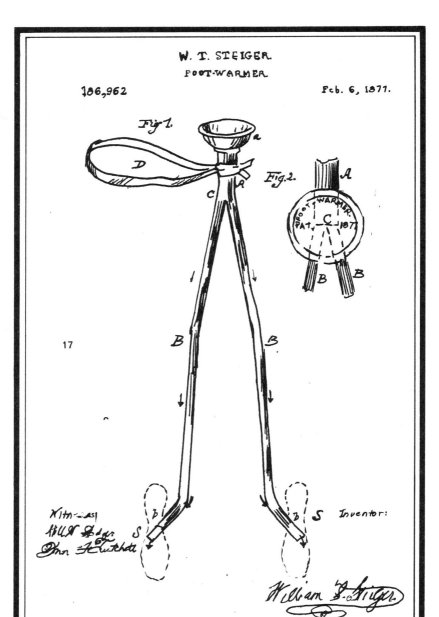

17. Foot-warmer. Original patent number 186,962, February 6, 1877. *Inventor: W.T. Steiger*

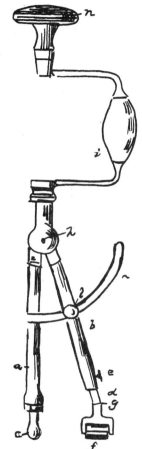

18. Device for producing dimples. Original patent number 560,351, May 19, 1896. *Inventor: Martin Goetze*

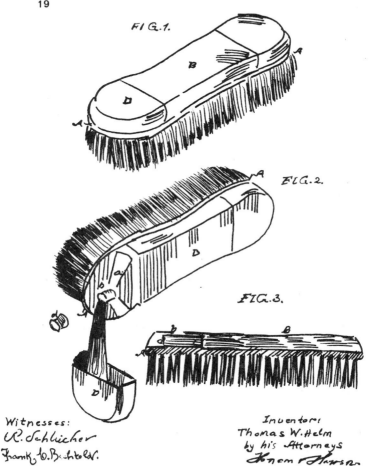

19. Combined clothes brush, flask, and drinking cup. Original patent number 490,964, January 31, 1893.
Inventor: Thomas W. Helm

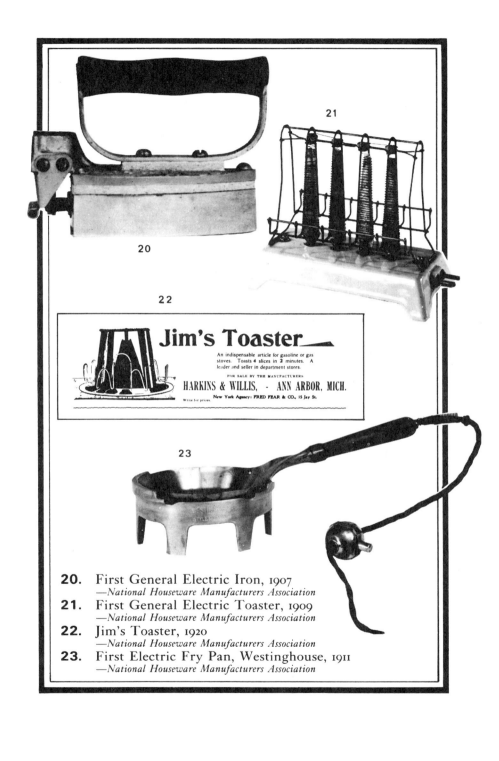

20. First General Electric Iron, 1907
 —*National Houseware Manufacturers Association*
21. First General Electric Toaster, 1909
 —*National Houseware Manufacturers Association*
22. Jim's Toaster, 1920
 —*National Houseware Manufacturers Association*
23. First Electric Fry Pan, Westinghouse, 1911
 —*National Houseware Manufacturers Association*

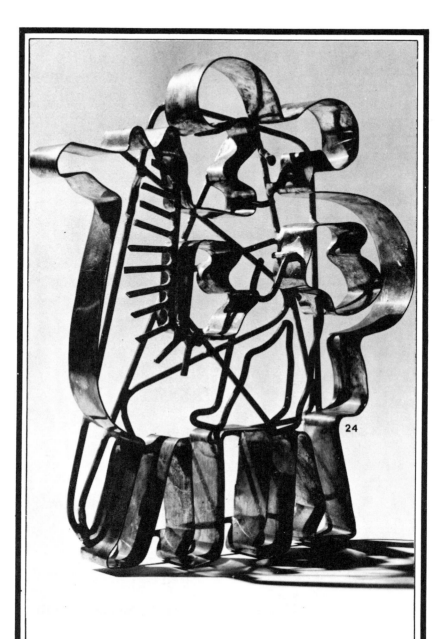

24. Lady Godiva Cooky Cutter
—*Smithsonian Institution*

BE IT EVER SO HUMBLE

Gadgets for the Home

Gadgets bought for the home tend to fall into three general categories; those that protect the homeowner from burglary and accidents, those that make household tasks easier, and those that help him save money, a fact of not inconsiderable importance.

Homeowners and apartment dwellers alike share one very serious fear—that their domicile will be broken into and burglarized in their absence. There are many gadgets available to hopefully deter the potential burglar, the most popular being the electric eyes and timers that automatically turn lights and radios on and off at random times, giving the appearance to an outsider that the house is presently occupied. Unfortunately, this is not the total solution, and as crime rises, so does our national investment in locks and alarm systems.

Some may think that our obsession with locks and keys is something new—a by-product of our current society. After all, haven't we all heard tales about the good old days when families left their home without ever thinking to lock the doors behind them? History doesn't bear this out. The oldest mechanical locking device was found near Ninevah four thousand years ago. The first keyhole was developed by the ancient Greeks, who also invented the first lock that made it possible to lock and unlock a door from either side. Locksmiths in the Middle Ages used things like hidden keyholes or studs placed on exterior decorations to foil would-be thieves. The first truly basic improvement in technical design, the lever tumbler lock, was introduced in the eighteenth century. It operated with a key that had to raise the lever to a specific height before the bolt could be removed.

Today's locks are much more complicated. Locksmiths would like to believe that their expensive models are "pick-free," but unfortunately that's not always the case. Luckily, there are alternatives. There is, for example, an inexpensive door chain that, though latched on the inside, locks from the outside with a special key. If a burglar does manage to pick the lock, the indoor chain is still in place, and hopefully he will think you are at home and quit while you're both ahead. Many people are partial to the various combination padlocks, which work well, but only to a point. There's always the possibility that you'll forget the combination yourself or that, by hit or miss, the burglar will discover it.

Perhaps the most innovative of the new locks is the Maglock, a lock without a keyhole. The "key" is a three-quarter-inch disc whose magnetic patterns align tiny magnets within the lock. The lock will open only if the magnetic "combination" is right. It seems to be more foolproof than most, but since no lock is really 100 percent, more and more people are backing themselves up with alarm systems.

One ultrasonic device, plugged into a light and/or alarm, will trigger the alarm and/or light when it "hears" someone enter the room and will continue activating these devices for two minutes after the person leaves. It then shuts, resets, and goes off again at once, in case the thief decides to give it one more try.

An electric alarm mat, activated by as little as fifteen pounds of pressure, will automatically turn on alarms, lights, buzzers, or sirens the moment an intruder steps on it.

Some favor the door jamb, which sends out a shrieking alarm if anyone attempts to force the door, and then can't be moved. The more pressure exerted, the tighter it holds and the louder the alarm.

If you happen to be home without these safeguards and your door is jimmied, it helps to have a chain lock that also doubles as a burglar alarm, which will both prevent the intruder from entering and hopefully scare him off at the same time.

Another gadget offering protection if you are at home when someone tries to break in is an electric siren alarm that attaches to doors and windows. You just pull the pin and it sets off a siren that will alert outsiders to the fact that you're in danger. The Invento alarm, a horn whose blast can be heard as far away as a mile, also serves the same purpose. All you have to do is pray that whoever hears your alarm or siren doesn't mistake it for normal city noises and ignore it!

The other fear urbanites and suburbanites share is that of fire. Often enough, fires are caused by short circuits, which can be prevented with very inexpensive gadgets like the Mini-Wonder, a safety-fuse circuit breaker that automatically shuts off electric current before a fuse is about to blow. No matter what the cause of the fire, it's reassuring to know there are special gadgets to protect you from serious injury.

One of these is a simple clocklike apparatus using no batteries or electric wiring that can be hung up near every fire hazard. The first heat of fire automatically sets off a warning alarm. Another is an all-purpose alarm that detects not only smoke and fire, but noxious or explosive fumes, cooking gas escaping from a blown-out pilot, and fuel-oil and gas leaks as well. Plugged into any 115-volt outlet, it operates via a thermostatic heat detector, gas sensor, and nonsparking buzzer.

Extremely valuable is the SmokeGard because just one unit can guard an eight-room house. It detects smoke and fumes before they can become lethal and has two other advantages. It can be easily installed by the homeowner himself, and since its power comes from batteries, there is no danger that burned-out wires will render it inoperable before it has a chance to sound off the alarm.

Once the alarm goes off, the next step, of course, is to get out of the house. No family should be without an instant safety ladder, which needs no previous installation. You just hook the top of it over any window sill and climb down the steel rungs to safety.

There are so many other gadgets available to safeguard the homeowner from other potential problems that I'd like to briefly mention a few of them.

A gadget called Dial-Temp lets you check the temperature of your home instantly from anywhere in the world. If it's too hot or too cold, you can notify a neighbor to make the necessary adjustment. Also, if you're away from home, there's a furnace alarm that plugs into a lamp that automatically lights up to alert an outsider when the heating system fails and the inside temperature drops to 55 degrees or below.

There are flashlights that send out Morse-code signals and key chains containing tiny flashlights that make it possible to find the keyhole quickly in the dark.

There are self-coiling power cords that adjust to the exact length you need and lessen the risk of accidentally tripping over the slack.

There are foot grips made of steel, adjustable to any size shoe, that ensure a sure grip on ice, snow, or slippery grass.

Cordless electric lights are useful for areas where no outlets are handy and may be vital when light is needed in an emergency.

Shock rods, weighing only twelve ounces, offer great protection against a mugger and are legal everywhere. Pushing a button releases twenty-five volts of electricity that will momentarily immobilize an intruder or an attacking animal, but will cause no injury. Tear-gas dispensers of various sorts are also available but are less widely acceptable legally. Check the laws of your state before you make the investment. And no matter how much you want to protect yourself, remember one thing: the "Saturday night special" is *not* a gadget.

An interesting device patented but not yet marketed should prove quite popular with apartment dwellers who live alone, especially the elderly. Called the Tenant's Emergency Indicator, it consists of a colored photo with seven flowers printed on it clockwise to represent the days of the week. The photo is inserted on the outside of the door into a clear plastic holder in the center of which is a pointer, pivotally connected to the top part of the holder. Each day the tenant moves the pointer so it points toward the flower symbolizing the specific day of the week. If the pointer isn't moved one day, this will alert a neighbor or the superintendent that a possible emergency may exist in the apartment.

Now let's look at some of the gadgets that make household tasks easier and more enjoyable. With a Mini-Breaker circuit protector, fuses never have to be replaced. If there is an overload and a short circuit does occur, all you do is press a button in the circuit box, and the electric service is restored.

For years, people have been enjoying remote control devices that allow them to turn on appliances, lights, TVs, and stereos within a specific distance. Now, RCA is researching a new remote-control switch, a solid-state device small enough to carry in the palm of the hand that will let you turn appliances on and off from anywhere in the house. An even more sophisticated version is being contemplated that will make it possible to turn on the heat and water at a distant vacation house before leaving your own house in the city.

An inexpensive Venetian-blind cleaner can wash four slots, both sides, all at once.

Various timers will turn appliances, lights, and stereos on or off automatically at any given time. Some have buzzers that can be preset twenty-four hours in advance for as many different times as you have to be reminded that you have something important to do. Pocket timers are helpful as a check on parking meters, long-distance phone calls, or cooking when you're out of the kitchen and can't hear the kitchen timer go off.

A stud finder called the Magnicator, operating, as its name indicates, by magnet, makes it easier to hang pictures or put up cabinets, because with it you drill or nail right into the stud instead of hitting empty space in the wall.

It used to be that a certain portion of the family budget would have to be set aside for home emergencies or jobs that would require outside professional help, such as plumbers, window cleaners, and painters. Today this portion can be reduced considerably. There are many gadgets one can buy that will pay for themselves a hundred times over by saving your having to go to outside sources. Take clogged drains, for example—a problem that invariably meant a call to the plumber, at eight dollars an hour. Now you can often solve the problem yourself. One way is to use a Mini-Jet type drain unclogger, which clears up most grease-, detergent-, or hair-clogged drains by turning ordinary water pressure into a powerful jet action that blasts away the obstructions and prevents them from backing up. New from Belgium is a 'syphon pump,' a hand-operated suction device guaranteed to unstop a drain even if the trouble spot is as much as thirty-five feet down the pipe. There are also gadgets that actually prevent clogging, such as the Hair Snare that is used in the shower or bathroom sink when taking a shampoo; it fits on top of the drain, and thanks to a fine screen, captures the hair before it has a chance to stop up the drain.

Window cleaners also cost money. A couple of years ago, no one in his right mind could see himself hanging outside a tenth-floor apartment window to clean the outside of it, no matter how much money it would save. Now, with a Magna-Clean, you can wash both sides at once—from the inside. It works like this: A damp cleaning tissue is placed on a magnetized hand unit, which is used on the inside. A steel plate with another damp tissue is placed on the outside. The magnet on the inside holds the plate on the outside. As they both move over the pane, the dirt is washed away on both sides at once. When finished, a cord is pulled to retrieve the outside plate. All very simple.

The cost of hiring house painters has skyrocketed, but luckily, there are a great variety of aids that make painting your own home much more economical and almost as professional. I'll mention a couple. One is an electric paint remover that can quickly strip away up to twelve layers of old paint, wallpaper, or tile, eliminating the endless hours of scraping and steaming them off. When this is completed, it is easy enough for even the most inexperienced painter to apply the new paint with either an automatic paint roller, which dispenses the paint directly on the wall, or the Electric Airless Paint Gun, which paints or sprays an even line from one-half inch to eighteen inches wide in one stroke. Both are guaranteed drip-free and do away with the need to clean up messy brushes and paint pails later.

With inventors coming up with one ingenious household gadget after another, it really may not be a fantasy to forecast a day, hopefully soon, when, except for extreme emergencies, having to depend on outside help will be a thing of the past.

THE GADGETING GOURMET

Gadgets for the Kitchen

Examining gadgets today in comparison with those of a hundred years ago leads to two rather diametrically opposed observations. The more things change, the more they stay the same and, the converse, the more things change, the more things change.

One reason why certain of the gadgets we own resemble those of our forefathers so closely is because, caught up in the wave of nostalgia, we are buying them specifically for that reason. Advertisements for such things as apple parers, wrought iron napkin holders, wooden pepper mills, cast iron nutcrackers, and manual coffee grinders all stress the word *old-fashioned*.

Another reason is that a hundred years ago gadgets did the jobs they were created to do—simply, yet well. Today, some of the jobs still need doing, and often, simplicity is of great value. It is hard to improve on a turn-of-the-century metal ball ricer, wooden crepe spreader, potato masher, chafing dish, wooden steak beater, potato-and-carrot serrators, cabbage graters, olive spoons with handles terminating in spears to lance the olives, three-tined pickle forks, not to mention the inventive dessert molds and tubes and ingenious pastry cutters.

On the other hand, there are a great number of gadgets that have been updated in ways that make them even more appealing. Grandma's old-fashioned rolling pin becomes an adjustable one with three sets of removable discs to permit rolling out the pastry in the exact thickness specified by the recipe: one-quarter inch for shortbread, one-eighth inch for quiches and tarts, and one-sixteenth inch for very thin pastries. And with this particular item, dough, for the first time, can be rolled onto the pin and unrolled directly over the dish.

The knife sharpener that in 1905 consisted of a can filled with sand with slots in the lid to slide the knife through back and forth becomes either an electric sharpener or a Dial-X sharpener that, with the twist of the dial, adjusts automatically to give the proper bevel and taper for every type of blade, guaranteeing a razor-keen edge.

The ice chopper of 1885 was shaped like a chisel. A tong or metal prong with four tines was attached to a wooden handle by a brass chipper. The owner stabbed this in and out of the ice until he chipped off the amount desired. How much easier it is today to press a button in a modern refrigerator and get exactly what you want with no effort at all.

Cheese graters and slicers have also grown up. The latest in slicers is a cheese wire/slicer from Germany. The owner slides the roller across the cheese, stops it when it reaches the point of thickness desired, and presses the wire down. To grate cheese in seconds, he just drops it into the electric cheese grater, and *voilà*, it's done.

Salt and pepper dispensers have also come a long way. One tiny combination salt-and-pepper set operates by pushbutton. Another is in the shape of a fish. A twist of the silvery tail brings a sprinkle of freshly ground pepper from its mouth. A no-clog salt and pepper set provides an ingenious solution to an old problem: the spikelike teeth in the hinged spring tops automatically clean and open the holes, keeping them moisture- and spill-proof. And finally, a two-in-one combination wooden pepper mill and salt shaker—it opens in the middle to fill and contains a precision grinder mechanism made of stainless steel; grind it and out comes pepper; turn it upside down and out shakes salt.

A comparison between the heavy wooden lemon squeezer of yesteryear and the devices for extracting juices today shows the world of difference electricity can make. The days of hand-squeezing oranges one by one over an inverted cone are becoming a memory. Today, a great number of electric juicers are available in combination with blenders and mixers. The two most popular models that exist for juicing alone are the Fresh Vitamin C Fruit Juicer and the Electric Vegetable Juicer. The first operates by the pressure of the fruit-half against the reamer. The motor stops when the fruit is lifted off. Then the stirrer attachment juices the small pieces of pulp by forcing them through the strainer. The latter is a powerful push-button extractor that electrically pulverizes fruits and vegetables at the same time that it separates the juice from the pulp and seeds.

The first grinder that served more than one function was the 1897 Universal Food Chopper, which ground vegetables as well as meat. Embellishment was added to embellishment, culminating in the many combination meat-chopper-and-salad-makers available today that electrically slice, chop, shred, julienne, and grate vegetables, fruits, meat, and fowl.

But even these pale in comparison with a new gadget I doubt our grandparents could ever have envisioned—the Ronson Foodmatic. It operates via a power module with a one-third-horsepower motor and mixes, blends, slices, shreds, grates, grinds meat, sharpens knives, juices oranges, crushes ice, grinds coffee, makes ice cream, and even cooks. It is curious to realize that if things change as much in the next hundred years as they have in the last, even something as extraordinary as this may well be looked back on as "just another little gadget."

An interesting direction in kitchen gadgets is reflected in the growing number of devices that hasten the actual preparation of foods and enhance the serving of them. A look at what can happen with an ordinary roast will illustrate. No longer does an average size roast have to take three to four hours to prepare. No longer must the cook buy expensive meat or use special condiments to guarantee that it will be tender when done. No longer need he spend time in the kitchen turning and basting the roast, checking to see if it is done, and wondering if it will come out the way he wants it to. This has all been changed with the help of gadgets. An adjustable roast rack that locks in five positions makes special turning unnecessary. Self-internal basters free anyone from the need for constant surveillance. Meat thermometers show the temperature of the meat at all times. Boeuf à la Mode needles are available for enriching lean meat by inserting strips of fat into it.

Once prepared for the oven, time and money can be saved by insertion of a gadget known as Thermo Pin or King Roast Pin, which reduces shrinkage by up to 50 percent and cuts the actual roasting time by 35 percent. It does this by heating the meat from the inside out, sealing in the juices so it is tender, tastier, and evenly cooked throughout. When heated from the outside in, which is what happens in a regular oven, meat tends to shrivel and toughen. A thermometer timer dial at the top of the pin sounds off and stops automatically when the desired degree of doneness is achieved.

Once done, a set of heavy-duty forks with curved tines especially designed to grip any size roast aids in the transfer of the meat from the oven to the serving platter without the danger of grease stains or burns. For those who prefer to slice the roast by hand, this is made much easier by the many attractive grip holders available, and, of course, for those who prefer, there are the many electric knives and slicers.

A parallel to the Thermo Pin for roasts is the Perfect Magic Arrow for steaks. This, too, cooks the meat from the inside out and has a thermometer that shows when the steak is ready. A more ingenious device, however, is the battery-operated Steakwatcher, a computer for broiling steaks. The desired degree of doneness is selected on one dial; the thickness of the steak is registered on another. The machine is turned on and the computer signals when the meat is ready to be turned over, sounding off once again when the steak is done. No margin for error.

Most fish and shellfish lovers own gadgets they are partial to. These may include fish poachers with removable racks, hinged fish grills that can be used upside down for more intense grilling, fish scalers for whole fish that eliminate the problem of flying scales, combination shrimp shellers and deveiners, lobster scissors, and the familiar clam openers and shuckers.

Fanciers of other foods have not been overlooked. There are meat presses and Swedish-meatball makers for those who might want them; poultry clip holders that eliminate tying or sewing chickens before roasting; Teflon-coated bacon grills that prevent curling, spattering, and shrinkage, drain fat, and pop open automatically when the bacon is finished; and steel chestnut pans with holes in the bottom for roasting chestnuts over an open fire. And for the gourmet, a duck press, to extract juices from a cooked wild duck for use as sauces, or a couscous cooker with which to prepare the traditional semolina dish of North Africa.

Let's take a look at some side dishes. Salad greens can be washed and dried by one of the many hand-turned spin salad driers or by a Swiss drier that removes the excess water from the lettuce by centrifugal force. Salad tossers are plentiful, some doubling as popcorn butterers or dressing mixers, while others include special repositories for grating lemon, garlic, and other herbs. Once ready to serve, combination fork-and-spoon tongs made of wood or lucite and handblown glass vials, one for vinegar within an outer vial for oil, add the perfect touch.

The baking time for potatoes can be cut in half with an aluminum rack on which four potatoes can be speared; once in the oven, the metal prongs speed the baking time by conducting the heat inside. The Spud Baker has, in addition, a looped handle to ease lifting it from the oven, and it folds flat for storage. An electric Tater-Baker, a stove-top oven, uses only one-twelfth the heat of a stove and can function as a crisper or bun warmer as well. And to ensure that the butter will spread evenly throughout the potato rather than stay put in the middle, there is a baked-potato puffer that punctures the potato in such a way that, as it is described in an ad, "your crusty morsel will burst open like a floral bouquet."

The Salton Hottray which keeps food warm without drying or further cooking it, is not new on the market. It was first introduced in 1948 but didn't catch on. The first year, sales came to only $12,500. It took another ten years before sales would hit the $1 million mark, and today it is a multimillion-dollar item. Beginning to rival it, however, is the Invento Prima Infra-Red Food Warmer, which thaws frozen foods as well as keeps food warm. Incidentally, for those who just want to keep their plates warm, there is a thermostatically controlled Electric Plate Warmer, which looks very much like an electric blanket with slots in it, that can heat ten plates in fifteen minutes.

Worthy of mention at this point are a few of the very inexpensive gadgets that make the preparation of food so much less complicated than it would be without them: a Boilmaster, a three-and-one-half-inch heavy-gauge stainless steel disc that sits on the bottom of a pan and prevents the boiling over of milk, water, soups, sauces, or any liquid; a metal heat diffuser, used over the stove, that spreads the heat evenly and prevents burning; an asbestos pad to be put under pottery or chinaware to protect it from an open flame; a punctured frypan cover that stops grease spatters but at the same time lets steam escape so that the food is fried crisply; and the thirty-inch Reach-It, the third arm that lets you reach up to the top shelf without a ladder and put back some of the above-mentioned gadgets that have helped so much to take the irk out of housework.

Walking through the housewares department of any major department store, one can't help but be impressed by the number of gadgets created to help you prepare one foodstuff in particular—the egg. Let's say you want to boil it; the first thing you'll want to know is whether or not it is fresh. Your great-grandmother held it in front of a candle and determined its age according to the shadowy shape of the yolk. You put it large-end-down in an Eggs-Ray, whose battery-operated light shows how big the bubble at the bottom is. The smaller it is, the fresher. The next step is to choose from one of the many piercers, punchers, and egg picks available that pierce the shell so that it won't crack while boiling. The egg timer is then set to the desired number of minutes. After removal from the stove, an Egg Thing, resembling a corkscrew, can be used to break off the top of the egg. Or else an egg scissor, whose sole function is to snip off the top of the egg, will accomplish the same thing. Finally, the egg is placed in a special egg cup and eaten with special little egg spoons.

Perhaps omelets are more to your liking. There are special omelet whippers to whip the eggs when you start and special aluminum omelet lifters for turning and removing the omelet from its very own teflon-lined turnover omelet pan.

For those who prefer their eggs poached, the Bazar Français sells individual insertable French egg poachers as well as stands that hold six of them. The Gourmet Egg Cook poaches four eggs using only two teaspoons of water in the center receptacle; the glass cover over the cast iron pan diffuses the steam needed to poach them. Another gadget is a three-cup pan insert that fits into a larger pan that is filled with the water that sets the eggs. And the most expensive one is the Electric Egg Cooker/Poacher that includes an automatic timer and temperature setting. It poaches up to four eggs, boils up to eight, and they can be served directly from it at the breakfast table.

Not to be overlooked are some of the other little egg

gadgets, such as the familiar coddlers, hard-boiled egg slicers, wedgers, and yolk separators.

Space doesn't permit the listing of all the gadgets on the market that one could buy, but suffice it to say that anyone who spends any amount of time at all in the kitchen is more than grateful for the help supplied by such friends as grapefruit knives and spoons, tomato slicers, pineapple corers, orange peelers, cherry pitters, vegetable mincers, scooper-scrapers, sugar dispensers, asparagus cookers, butter cutters, snail holders, potato curlers, onion holders, radish rosers, garlic presses, spaghetti forks, parsley mincers, onion choppers, muffin breakers, vegetable steamers, stir and sip spoons, French-fry-potato cutters, and gravy separators.

This chapter wouldn't be complete without a mention of some of the more interesting gadgets that have been recently introduced or improved upon. The Hot Dogger introduced in 1919 has gone electric and now cooks six hot dogs in just sixty seconds.

For the hamburger lover, trays have been developed that shape, store, and dispense them on much the same principle as the ice-cube tray. Called the Pop-It Burger, it eliminates the messy hand-shaping of patties and the individual wrapping and unwrapping. It has four molds to hold the ground meat, which is pressed down with a knife or spatula. The trays seal airtight when stacked, and a twist of the tray releases the meat when it is removed from the freezer. Another hamburger helper is the Patti-Stacker, an acrylic-plastic set with eight separate discs with which one can separate and stack up to ten four-inch patties. For some, a hamburger is not a hamburger without ketchup. For them there is a handy Ketchup Saver, which lets you control the ketchup flow so you get only the amount you want, no more, no less.

An electrically dehumidified brisker that keeps crackers, cookies, cereals, nuts, pretzels, and potato chips crisp and fresh is being touted as the gadget "sleeper of the year" at Hammacher Schlemmer.

The Thermo-Spoon measures the temperature of the liquid in the bowl of the spoon, wonderful for mothers who are feeding their infants.

A butter spreader-dispenser is good for children, who can now butter their own corn on the cob or toast without getting the butter all over everything else at the same time.

A Liqui-Pour makes it possible to open, pour, and store fruit juice right in the can. The can opener becomes a pour spout that seals airtight for storage when it is recapped.

An English silver-plated toast warmer holds four slices of toast, while the candle-warmer keeps them hot. And the Slim-A-Slice cuts a piece of toast to the precise thinness desired.

Hoffritz sells a miniature individual tea infuser, and the Electric Teasmade first boils the water and then automatically pours it into a four-cup pot. A buzzer tells when the tea is ready, a soft light indicates that it is still hot, and the clock tells what time it is when you drink it. And the Standard Electric Kettle boils a pint of cold water in less than two minutes, holds two quarts, and features an automatic safety thermostat, easy-pour spout, and prongs to protect the table top when it is placed on it.

There are scales that show the ounce-gram measure at the same time, very helpful to dieters and those who want to convert European recipes. There's also a weight ladle, on one end of which the number of grams or ounces desired is calibrated; the bowl at the opposite end is filled until the balance levels.

The Home Potato Peeler electrically peels up to one and one-half pounds of potatoes in less than three minutes, converting the peels into foam so they can be rinsed down the kitchen drain.

The Flavor Jector makes it possible to inject whatever flavor you desire directly into meat, including garlic, mint, onion, wine, or chocolate mousse.

The Electric Food Smoker can smoke and cure cheese, fowl, meat, or fish in four to twelve hours and twenty-five pounds of meat in twenty-four hours.

And the Tri-Fry pan lets you cook three different foods at the same time. Teflon coated, it keeps the greasy foods separated from the dry ones, lets foods retain their separate flavors, and saves time when it comes to cleaning up.

Ah yes, cleaning up—which reminds me of one of the drawbacks to gadgets—they can't do everything.

I GET HIGH WITH A LITTLE HELP FROM MY FRIENDS

Gadgets for the Bar

Gadgets for the bar have an endless fascination for those who buy them. They are truly gadgets in every sense of the word. They serve a function, invariably they are unusual, they are fun, they serve as status symbols, and they are wonderful conversation openers.

Just as every cook is looking for the gadget to guarantee the "perfect cup of coffee," every bartender, however amateur, is searching for the one to guarantee the "perfect martini." Since everyone's taste is different, this is a fairly impossible goal, but, thanks to the various devices available today, it certainly isn't because nobody has tried, hangovers notwithstanding.

53 • I Get High with a Little Help from My Friends

For some, the perfect martini is made with 5 parts gin to one part dry vermouth. Yet there are those who won't touch it if the ratio is anything less than 8 to 1. But would you believe 724 to 1? You can now buy a martini dropper calibrated to 724 drops. You select the ratio of gin to vermouth you want, put the liquor in the tube as required, and press away, drop by drop by drop, anywhere up to 724. Then there is the Scientific Martini Maker with which the exact amount of vermouth is calculated to the desired ratio and added "scientifically" to the gin via an attached bulb that squeezes it out.

Some drinkers have even taken to using a gadget that was originally developed to spray mist over plants to spray vermouth over the gin. And for those who can afford it, there is the Electric Cocktailmatic. The gin or vodka and dry-vermouth bottles are installed on top. The host sets the ratio selector as far as he wants up to the ratio of ten to one, pushes the button, and the precise amount is automatically released into a measuring cup—the ultraperfect martini made to order.

For many, a cocktail is not a cocktail unless it contains ice. Many modern refrigerators contain special ice makers and refill units and there are an incredible number of individual models that make cubes in different sizes and shapes as well as fine, medium, or coarse crushed ice in a matter of seconds.

One of the most simple, yet welcome, gadgets involving ice is the Spil-Gard ice tray, which has a very interesting story behind it. It was created by Dominic Tampone of Hammacher Schlemmer, who was frustrated by the fact that whenever he refilled an ice tray, some of the water spilled over as he carried it from the sink to the refrigerator. This problem is not new to anyone who has ever filled an ice tray, but Mr. Tampone was in a position to do something about it. He went to the men who designed the ice-cube trays at Frigidaire and discussed with them the problem and what he saw as the solution, a special tray with side vents that would catch the water on either side whenever it was in motion. They, in turn, invested close to $2 million researching it, which shows how much money can go into developing even the least complicated of gadgets. They finally decided it was feasible on a mass-market basis, and as a result, drip-guard ice-cube trays and their variations are included in all new refrigerators and are available quite inexpensively in houseware departments and hardware stores everywhere.

I might add one thing for "the other half," those who not only don't like ice in their drink but, on occasion, actually prefer some drinks warm: there is now a gadget in the shape of a metal tube that, when put in cold beer, warms it up to "fashionable English tepid."

Each year more and more people are developing an interest in, and very real appreciation of, wine. More and more books are being published on the subject. More and more people are registering for classes and seminars devoted to learning about wines, and wine-tasting parties are becoming more and more popular as a way of entertaining old, and making new, friends. As a result, more and more gadgets are being devised to bring out the sommelier in all of us.

The first step toward enjoying a bottle of wine, and often the clumsiest, is opening it. Curiously enough, the first corkscrew, the Walker Universal Self-Puller, was patented in 1909, and the basic concept of it—the familiar wire spiral screw with a crown and aluminum stopper attached to a small round metal bar fit into a wooden handle—did not change very much until the last five years. But even with the newer models, the one problem with the corkscrew is that there is always the chance that some of the loose cork will end up in the bottle, which is one reason, incidentally, that the first wine of the bottle is served to the host rather than the guests.

Now there are new devices, the most notable being the Corklifter and the Cork Pop, that have solved this. With the Corklifter, you insert a pressurized needle through the cork, pump a small handle a few times, and the cork eases out without breaking. With the Cork Pop, you again thrust a needle through the cork, but instead of pumping a handle, you merely press a capsule that releases enough gas to push out the unbroken cork.

Almost everyone has had the embarrassing experience of pouring the initial glass of wine, only to discover that a few drops have spilled down the side to stain the tablecloth. This need happen no more. There are a number of drip guards on the market today. One of the most interesting is the Dripless Wine Pourer, which is inserted into the bottle once the cork is removed. It has a double lip that ensures against spilling, because when the extra drops fall on the first lip, they automatically go into the second channel and are returned right back into the bottle. Another guard is in the form of a decorative sponge, which fits around the neck of the bottle and absorbs any excess that might spill down.

For those who serve white wine throughout the meal, keeping it cool at tableside over a long period of time can be a problem. Now there are a variety of small frigidaria (cold-energy liners that retain the cold and can chill bottles for hours without ice) and very attractive wine coolers with special tubes that hold ice and cool the wine as long as necessary.

Every host knows how difficult it is to reinsert the original cork in a bottle once it has been removed. So if you think you'll have any wine left over at the end of an evening, it pays to have a few of the adjustable corks on the market, which fit any size bottle on hand.

The most festive of wines is, of course, champagne. Sparkling wines have uncorking problems just the opposite of still wines. In their case, the cork is often a bit too ready to leave the bottle. Champagne acts as its own cork pop. When the pressure is released, the wine invariably bubbles off, and it's not uncommon for a few precious drops to overflow. It is difficult to believe that anyone could object to the joyous pop or really mourn the loss of a few bubbles, but, should that be the case, he is hereby advised that he can buy a Champagne Cork Tapper, an eleven-inch chrome-plated corkscrew that pierces the cork and is twisted down until the end reaches the bottom of the bottle. The top is a short twist-on, twist-off spigot that eliminates all of the mess—and most of the fun. And if any champagne is left over (Horrors!), there are specially designed recorks that will keep the remainder from going flat.

In this age of ecology, we are often concerned with what to do with the bottles once they are emptied. Hobbyists suggest buying a GlassMaker as a fun way of recycling them. This is an aluminum gadget with which one can learn to transmute bottles into decorative glasses, ashtrays, vases, or even candle holders.

Most homes won't qualify for Frankfurt, Germany's contribution to ecology, but if it does happen that you are interested in pulverizing up to 1,500 bottles an hour, the new Breakmatic is something you should definitely look into.

As the party comes to a close, what better way to end a pleasant evening than to share a warm fine brandy with friends. For this, all you need do is take out your special brandy warmer that includes a brass stand, alcohol burner, and twenty-two-ounce snifter. Add the brandy, light the burner, and offer a toast to health and happiness. Cheers!

And to the genius who comes up with a gadget that ends hangovers, I propose a personal toast—and a question: Where were you when I needed you?

AN OUNCE OF PREVENTION

Gadgets for Your Health

People have always been concerned about their health, but never before have there been so many commercial devices available to the layman that can help prevent, diagnose, measure the degree of, and sometimes control certain illnesses or unhealthy habits.

Ever since the U.S. Surgeon General's office issued a report ten years ago that linked the smoking of cigarettes to cancer, many smokers have tried to kick the habit. Among those who have been successful, many credit a special Dial-Down cigarette holder with making it easier, because it let them taper off over a period of a few weeks rather than cut down "cold turkey." The secret is in the scientifically designed dial. When set at 100, 100-percent pure smoke is inhaled. Set it down to 80, and there's 80-percent smoke, 20-percent fresh air. The lower you dial as the weeks go by, the less smoke and the more air you inhale, until finally you've dialed down to zero—no smoke at all. And you've cured your habit.

Many who find enjoyment in the sheer physical act of smoking but want to avoid the dangerous side effects swear by Aquafilters, which, they claim, reduce by more than 75 percent the tar and nicotine content normally inhaled without altering the taste.

To cut down on his smoking, Soviet party leader Leonid Brezhnev carries a cigarette case with an automatic timer lock that allows him only one filter-tip cigarette an hour.

Many men have turned to pipes as an alternative to cigarettes, since they are less lethal. One of the most ingenious is a small pocket model from Brussels. It is structured so that the filter won't allow any nicotine juice or tobacco to reach the mouth. It smokes dry and also has a mobile grille cover that prevents hot ashes or burning tobacco from falling out and burning holes on the table or rug.

Burning ashes from cigarettes and pipes on clothing, home furnishings, and especially bedding have always been a cause of concern, if not downright alarm. A number of gadgets have been devised to prevent this. One of the most popular is the Safety Smoker Robot, which secures the cigarette at the same time that it automatically collects the ashes. Another interesting device patented but not yet marketed doesn't exactly double as an ashtray but does serve as a preventive against fires caused by dropped or lost cigarettes: it is a cigarette holder-clamp, shaped much like a clothespin, with a magnetic base that attaches to any metal surface. The clamp is firm enough to hold the cigarette securely without putting it out, can be moved around at will, and can be suspended over any ashtray.

Many pipe smokers are partial to a special ashtray that not only collects ashes but cleans the pipe at the same time. The pipe is pushed over spring-steel blades that draw the ashes out automatically and drop them into the tray. It not only prevents the residue ash from building up in the pipe, it also permits the moisture in the stem to dry out more easily. And lastly, the newest addition on the market is called Hank's Pipe Cleaner, the first to remove tar and nicotine by means of a tiny portable steam machine.

Since cigarette smokers are bound to smoke no matter what graffiti the Surgeon General puts on their cigarette packs or what laws are passed to segregate them, let's take a quick look at some of the gadgets they use.

Cigarette lighters have long been objects of fascination. Not only do they serve a purpose, but they are probably the most popular gift item in the world. They come in every shape, price, size, and design and in combination with things ranging from key rings and pencil tops to watches and umbrellas. Speaking of umbrellas, you can now attach a dispenser that holds five cigarettes to it. If you want to smoke while you're walking in the rain, all you have to do is push a button that drops one cigarette out of a doorlike compartment near the handle onto a tiny pulley that carries it to another thin pulley that simultaneously ignites it. All that's left for you to do is inhale.

One problem with lighters is that all too often, and when least expected, the flints wear out. One alternative is an electric metal lighter that, via batteries, ignites the cigarette instantly as soon as it is inserted and then pulled out of the lighter. Another interesting gadget is the solar cigarette lighter, which also operates without fuel or flame. Its inventors claim that when the cigarette is placed in a special holder and the lighter is pointed toward the sun, it will light automatically. But the best solution is probably found in Ronson's new Electronic 7 battery-propelled lighter. When the automatic switch button is pressed, it discharges the capacitor, in which a 15-volt battery is stored. The tiny transformer instantly boosts the voltage and sparks it across the butane burner, which has simultaneously emitted gas to be ignited. The battery is said to provide up to 25,000 lights. Another interesting feature is the lighter's flame adjustor, which allows the smoker to preset the butane height: low for cigarettes, medium for cigars, and high for pipes.

What may be an indicator of the lighter—or nonlighter—of tomorrow was displayed at the 1973 International Patent Licensing Exposition by Kyu Bong Whang of Seoul, South Korea. To illustrate his invention, he pulled out a pack of his special cigarettes, each of which had a 4-mm. odorless, nontoxic "igniter" on the end. A cigarette was then brushed lightly across a chemically treated strip on the package and flared immediately into flame.

And the future? Science fiction writers predict we'll soon have cigarettes that are specially treated so they burst into flame at the first puff. No more matches, no more lighters, and, alas, no more thoughtful pauses.

Heart disease is one of the country's leading cripplers, and as we learn more and more about what causes it, we also learn what we can do to prevent it.

Most doctors agree that one important way to prevent heart disease is through exercise. Americans, especially the urban American, just don't get enough of it. Interestingly enough, the exercise apparatus was extremely popular in the late 1800s. The Sears Roebuck catalog of 1885 has pages and pages of advertisements for hydraulic rowing machines, wooden and nickel-plated dumbbells, striking bags, and even portable home gyms. The same equipment, in more sophisticated form, is just as popular today, as are some innovations our forefathers didn't know about or perhaps didn't need as much as we do now.

One is the Executive Jogger, for those who really don't have time for that quick jog around the park before going to work. It consists of a foam-cushioned pad with springs inside, slightly slanted to emulate uphill running. Six minutes a day on this is equal to a mile of outdoor running.

The Pocket Grip Exerciser, using the principles of isometric tension, is good for building up muscles in the wrist, grip, and arm.

An eleven-inch Peda-cycle, which can be put under the desk and used while working in the office as well as when at home watching TV or reading a book, is a boon to the lazy person. As an exerciser, it is more convenient, more relaxing, and less strenuous than bike riding, yet the stomach, thigh, and leg muscles derive the same benefits. It has a scientific tension control so you can set your own pace and has extra value because it can also be worked with the hands to firm up the arms, shoulders, and chest.

Many major companies now have fully developed health plants in their buildings and require that their executives participate in a physical fitness program. Mobil Oil recently ordered motorized treadmills (at $3,000 apiece) along with electronic pacers that synchronize time and exertion so the executives can pace themselves correctly by watching and matching up needles on the dial faces. One executive was quoted as saying, "When you are making $50,000 a year, you don't grab the oars and start rowing while muttering, 'And a-one, and a-two, and a-three. . . .' You have a gadget that counts for you."

One of the problems originally facing the astronauts was how to keep physically fit during their long journeys into space. After much experimentation, a special isometric/isotonic, light, portable, compact machine that exercises the entire body was developed and is now available to the general public for use at home or in the office.

Very recently a patent was taken out on an exercise bicycle that claims it is the only one that provides a person with not only both leg and arm activity but body massage as well. As the bicycle is pedaled, air-filled rubber balls mounted on a belt roll around the waist. It is the first time, supposedly, one can experience physical stimulation and physical relaxation at the same time.

We all know that exercise is important, but equally important for psychological as well as physical well-being is posture. Therefore, it is not surprising that someone would come up with a gadget called the Belly Beeper to make sure we don't forget to throw our shoulders back and tuck our stomach in. The beeper is attached to a belt fit to size when you stand up straight. The minute you slouch or let your stomach out, you set it off and start hearing beeps that make you aware that you are doing something wrong. The only way to stop the beeps is to straighten up. Following the theory of behavior modification, you learn to tense your muscles on signal and supposedly get to the point where your muscles unconsciously anticipate the signals and tighten up before you have the chance to droop.

For those who have not been able to avoid heart disease and hypertension, there are devices on the market to make living with them as comfortable as possible. Some who have high blood pressure are advised by their physicians to check their own pressure at certain intervals. Fortunately, thanks to a portable Sphygmostat Electric Blood-Pressure Monitor that weighs less than three pounds and doesn't require the use of a stethoscope, this needn't be a major problem.

Those with cardiovascular problems or respiratory disease whose medical situation requires that they have a supply of oxygen on hand at all times can be grateful for the new, relatively inexpensive, lightweight, portable units by Dow Chemical called SOS, which stands for Solid Oxygen System. Available in two sizes, one weighs less than three pounds, is only ten inches high, and supplies fifteen minutes of oxygen at six liters per minute. The other is the size of a pair of binoculars, weighs four pounds, and supplies forty-five minutes of oxygen. The mask is attached to canisters about the size of a can of juice concentrate. Inside the canister, pure oxygen and water vapor are produced when solid sodium chlorate is reacted with a catalyst. The system is activated by simply pushing a button. Needless to add, these units are extremely good to have around in case of emergencies, such as fire or health accidents.

Doctors have always faced a difficult task in deciding just how much exercise is right for a cardiac patient. A team of German researchers have come up with a device called the

Cardiomed that monitors the working heart and tells when it is not beating at the proper rate. Not much larger than a billfold, this battery-operated gadget checks on the heart through electrodes attached to the chest. It emits a single beep whenever the heart rate falls below a predetermined unhealthily low level, a double beep whenever it rises above a preset ceiling, and is a valuable asset in getting a cardiac patient started on a postattack exercise program.

Of course, one of the most important gadgets for the heart, and one that has deservedly received quite a bit of attention recently, is the four-ounce nuclear-powered Pacemaker, which can last up to ten years, a marked improvement over the original ones whose batteries had to be exchanged every year or two.

Dentists are increasingly stressing the importance of preventive dentistry, and as a result, we are paying more and more attention to our teeth. Water-piks have become a way of life, as have cordless electric toothbrushes. One company recently introduced a hydraulic dental appliance that attaches to any faucet. It operates with a water jet, the pressure of which is controlled by an on-off button, and gets into areas like crevices and under the gums that the ordinary toothbrush just can't reach. In addition to stimulating the gums and flushing out disease-causing food particles, it has another advantage in that the owner can control the water temperature if he has sensitive teeth.

Still in the experimental stage, but hopefully available soon, is a spring-loaded electrically powered toothbrush whose bristles will automatically operate up and down, as dentists recommend, as well as from left to right, which is the way most of us tend to brush.

Those who've paid the price for neglect of their teeth with dentures are finding the going a bit easier with Dent-U-Sonic, a device that cleans removable dentures electrically with chemicals and sonic vibrations, eliminating the embarrassing overnight soaking.

Since many gadgets are created to make things easier for

the owner, it should come as no surprise that a number of them have something to do with getting toothpaste out of the tube. One is a push-button dispenser that attaches to the bathroom wall and when pushed, lets out just the right amount. Another is an automatic dispenser operated by a key. Still another eliminates waste by inserting the tube into a plastic jacket; when the slide is pushed, the toothpaste is automatically forced up from the very bottom. Both of these prevent the toothpaste from hardening at the top. And a new idea is the capless toothpaste tube, which automatically dispenses enough for one application on the brush and seals instantly when the squeezing stops.

There's a Sip-Up toothbrush that works like a straw for easy rinsing. To rinse out the toothpaste, just flip the brush over, place the handle under the faucet, sip up the water, swish around the mouth, and expectorate. Next, a new brush that contains its own supply of toothpaste directly in the handle. The user opens a slide, turns the knob, and out the toothpaste comes directly onto the bristles.

What is left? Fred Borch, retired chairman of General Electric says, "Maybe we ought to consider making a combination toothbrush and shoe polisher for the man who, whenever he opens his mouth, puts his foot into it."

One could devote an entire book to gadgets that have been developed to make man feel more comfortable in relation to his body. I'll mention just a few.

Hearing aids are accepted as commonplace, but carrying them one step further, a University of Utah scientist has just announced the development of an "artificial ear," a tiny microphone and subminiature electronic computer that is connected with the auditory areas of the brain and allows those who are totally deaf to hear rudimentary sounds.

For people such as construction workers whose work exposes them to a great deal of noise, there are Sonic Ear-Valves that filter out disturbing and dangerous high-frequency sounds while still letting through the low frequencies needed for understanding speech.

There are many aids for the insomniac, but few as interesting as the sleep sound devices that screen out disturbing noises and induce relaxation and sleep by lulling you into a vacuum of scientifically blended no-noise "white sound" rhythms or sounds that imitate the surf or rain. If the disturbing noise is snoring from the other side of the bed, however, more direct action is available. Wake him or her up and give him a pair of plastic Naso-Vents, which fit in the nostrils, open the nasal passages, and permit easier breathing. And if you *still* can't get to sleep, try reading a boring book (not this one) using one of the available flashlight attachments that fasten on to your reading glasses.

Disposable thermometers are not all that new, but the electronic thermometer that, when put to any point on the body, can give a reading in seven seconds is. And any parent of a restless child can tell you what a blessing that is.

Those who take out their contact lenses from time to time during the day can now store them in a special cushioned container that attaches over a watchband or bracelet.

Those into bio-feedback can now monitor their own brain-wave activity, heart rate, and skin-resistance feedback with a two-pound portable bio-feedback trainer.

Paralysis can be one of the most depressing and debilitating of physical illnesses, as George Wallace, governor of Alabama, knows. He was recently fitted with a device that can finally control his pain. Referred to as the TNS, it is a rectangular electrode the size of a nickel, which in his case was pasted to the skin overlying the spinal cord below the waist. Wires connect the electrode to a battery-powered transmitter the size of a hearing aid, which is attached to the wheelchair. Whenever he feels a twinge of pain, he turns a knob that sends up to ninety milliamps into the spinal cord, which produces a tingling sensation and, more important, cuts off the pain. "I feel," said the governor recently, "that this little gadget is going to do me some good."

Which all goes to show that all gadgets, contrary to what some cynics may lead you to believe, are not necessarily frivolous. Some day, our very lives may depend on them.

I GET BY WITH A LITTLE HELP FROM MY FRIENDS

Gadgets for Grooming

In 1973 alone, over $1.2 billion was spent on "personal care" products, a goodly percentage of this unquestionably on gadgets. To get a better picture of the variety and choice available, let's follow a mythical couple as they prepare to meet the day.

In the past, people considered bathtubs and showers well worth the expense and trouble of manipulating them if they just did their job—helped you bathe. Today, however, they reflect a certain life-style. Oftentimes they are equipped with gadgets that ensure that one is not only washed, but massaged, pampered, scrubbed, and relaxed as well. One such item, the AquaMassage Shower Head, releases alternating jets of water against the body "like the probing fingers of your own personal masseur. The water streams caressingly over the body like a waterfall. To get an entirely different feeling, just rotate the shower head." If more stimulation is desired, there is the Shower Mate, a nylon-bristled rotary brush that attaches to the shower wall and promises not only to promote circulation, relax muscles, and scrub, but to serve as a personal back massager and scratcher as well.

Those who prefer a bath can enjoy whirlpool hydrotherapy with an inexpensive attachment called the Aquajet. Fitting right into the bathtub faucet, it "lets the swirling water massage away tensions, tiredness, and muscle aches and pains while it gently stimulates circulation and soothes the entire body."

For the future, Westinghouse is experimenting with a "people washer," a shower system that will spray you with soapy water, rinse you off, and then dry you with a current of hot air. In addition, you'll be able to set the time and temperature according to your personal requirements.

But back to today. If washing the hair is part of the morning ritual, there is a way to do it without wetting any other part of the body. The Shower-All, a three-position special hose and hand implement that attaches to the shower, lets you direct the water to the exact place on the body you want it to go.

A favorite with the elderly and infirm is a bath security rail, attached to the side of the tub, that serves as an aid for getting in and out of a slippery bath. And to make it easier for the one who has to clean up, there is a no-stoop Tub Scrubber, a cellulose sponge head on a twenty-six-inch handle that lets one clean the tub while standing up.

Once out of the tub or shower, a moment of luxury can be enjoyed by drying off with towels warmed by an electrically heated towel stand, which also dries them out after use.

The man's next step is to shave. The first recorded shaver was a crude flat razor made by the Peruvian Indians over four thousand years ago. This evolved over the centuries into the straight razor, strips of hardened and tempered steel with hollow-ground edges held together by a pivotal handle. The major drawback of the straight razor was that it not only nicked the skin more frequently than not, it also required frequent stropping and honing in order to keep the edges keen.

It wasn't until 1913 that a new kind of razor was introduced to the general public. It was called the safety razor and was made by placing a metal bar between the blade and the area to be shaved and was constructed to prevent the deep cuts so often caused by the straight model. Today, with many improvements, such as double-edged blades, more comfortable holders, and cartridges, the safety razor is the most popular instrument of shaving in the world today.

The first electric shaver, introduced by Schick in 1931, was developed in an attempt to eliminate the lather and disposable blades that are part and parcel of the safety version. But even with these conveniences, plus the fact that they can be used wherever there is an outlet, even the cigarette-lighter outlet in the car, they have never been the outstanding sales success their creators forecast they would be. On comparison, men prefer the closer shave the safety razor gives them.

One of the newest and most innovative razors on the scene today is General Precision Corporation's Stahly Razor. Permanently lubricated and working without benefit of electricity or batteries, it lets an ordinary blade make up to eight thousand lateral cutting strokes a minute, all by means of oscillating vibrations.

Gadgets have also left their imprint on tasks associated with shaving. No longer do blades have to be discarded after two or three uses. A Span-o-Matic razor-blade sharpener makes it possible to get hundreds of shaves from the same blade. And men no longer have to suffer the initial shock of cold lather on their faces, because a special gadget can be added to the shaving-cream dispenser to warm it before it is applied to the brush.

One possible reason that the safety razor wasn't developed earlier than it was might be because so many men wore beards and mustaches years ago. There was never a great need for an instrument that gave a very close shave. That fashion is in vogue again today, of course, and with it, improvements on the gadgets of old, such as the mustache combs and curlers, and some new additions, like the electric beard-and-mustache groomer that comes equipped with a styling comb, barber comb, scissors, and styling razor, or the small mustache razor that holds a tapered blade with a five-eighths-inch cutting edge on one end and a one-quarter-inch edge on the other. Mustache cups are also making a comeback, with one variation. Now those in need can find a special cup for lefties that has the handle on the left and a specially fitted insert that fits over the mustached area and leaves the rest of the opening free.

The biggest advance in women's grooming came during the last decade with the introduction of the many lightweight, portable, relatively inexpensive electric setting, styling, and drying devices for the hair. Heated metal curling devices to "iron" or shape men's and women's hair aren't new. They go back as far as Biblical days, when they were used for religious occasions and special ceremonies.

Shortly after World War I, a number of electric curling irons found their way onto the market. The most popular set included a rod, cord set, curler clamp, and drying comb. The clamps and comb had to be warmed over the heated rod before they could be applied to the hair. This was the forerunner of today's electric comb. The one we use today was first introduced in Switzerland in the early 1960s, but the market for it in America didn't grow until, ironically, men's hair did in 1967. The hot comb soon followed in 1969.

The popular electric hot rollers that let a woman comb out her hair in ten minutes, putting an end to sleeping with curlers, were popular in Denmark long before they caught on here. The same holds true for many of the electric misters, spray stylers, and hair setters that exist today.

Portable electric hair dryers though not as complex as the

ones on the market today, which come in combination with brushes, combs, misters, special settings for wigs, and scalp stimulators, have been around for at least sixty years. The first electric hair dryer was actually the vacuum cleaner; clever women used the air from the exhaust to dry their tresses. The first commercial electric dryer was introduced in 1920, but it was too cumbersome to make much of an impression. Finally, in 1956, a table-top blower unit with a hose attached to a bonnet that fitted over the head caught the public's interest. These bonnet dryers lost their appeal in 1963 when the hardhat beauty-parlor type dryers were reintroduced along with the Salonette Beauty Mist dryers that boasted a built-in facial sauna. Two years later the hair "blowers" from Scandinavia made their mark, and today there is no end to the choices available.

Once the hair is ready to be combed out (more often than not with a Gillette Detangler that, via two vibrating combs, gets the tangles out of wet or dry hair painlessly) and it's time to apply makeup, a mirror becomes a woman's best friend. The first electric makeup mirror was marketed in the early 1930s. In 1966, an illuminated version that could be carried around in a compact travel bag took the country by storm. Other embellishments were added, culminating in an electric mirror with multiple light settings that could be switched to let the woman see how her makeup would look in daylight, during the evening, at the office, or under flourescent light.

As popular as electric mirrors are, hand-held mirrors are even more so. The ones that attract people most are the extension mirror, attached to a wall and magnified for closeup use, which can be adjusted to various angles; the hindsight coiffure mirror, extending thirty-six inches, which can be raised, lowered, swiveled, tilted, or turned in any direction; the neck mirror, which offers a perfect view of the back of the head, leaving both hands free to tease or comb the hair; and the adjustable standing floor mirror, which can be moved around to catch the best available light and can be stored in the closet if space is limited.

After making-up, many women are beginning to use a

new gadget just introduced in the United States, the Quicklime battery-run nail file, which quickly files the nails of the fingers and toes without chipping the polish, evens off and shapes them, and can also flatten corns and calluses.

The gentleman is ready to get dressed. If the crease has gone out of his trousers overnight, it is no problem as long as he has a Touch-up Steam/Press Valet, a two-pound gadget that, by applying the proper pressure and heat, brings the crease right back. If wrinkles are a problem, the electric Steamette will steam them away without ironing. This gadget can also be used to restore the nap to furs. And if the pants aren't pressed "just so," the Electric Pants Presser Valet will quickly take care of not only them, but jackets, ties, belts, and accessories as well.

The next step is to select a tie—by gadget, of course. The Cordless Electric Tie Rack, holding thirty-six ties that can be rotated by push button, simplifies the much more complicated process of handling and looking through thirty-six individual ties.

If you've ever had the frustrating experience of not being able to find the other sock, pick up something called the Pair-O-Soks, a tiny plastic ring that keeps a pair of socks together through the washer, dryer, and in the drawer until they're ready for your feet.

And now to shoes. If they need polishing, there are things like an electric Boot-n-Shoe brush, which cleans with plastic bristles, or a push-button shoe polisher with two separate bonnets, one for light, the other for dark shoes. And to help get the shoe on the right foot there is a five-inch telescopic shoe horn that extends to nineteen inches and simplifies slipping on the shoe without having to bend or stoop, a lifesaver for the arthritic, or the very lazy.

For those who prefer boots, there are boot shapers, which prevent boot odors, cracks, or creases, and boot huskers, which simplify the removal of boots by supplying a form-fit "something" to force them against.

And so, tressed and dressed, one is prepared to meet the day. Onward—to the office.

7

HOW TO SUCCEED IN BUSINESS WITHOUT REALLY TRYING

Gadgets for the Office

I'm sure it comes as no surprise that many people spend as much, if not more, of their time in the office as they do at home. And just as they turn to time- and energy-saving devices to simplify their life at home, they do the same at the office. And it was always thus.

Let's look at how our forefathers approached the situation at the turn of the century. The typewriter had originally been patented in 1829, but even by the early 1900's it was a far cry from the slick, efficient models we are accustomed to today. Today, we have typewriters that type complete sentences automatically, typewriters that type in nine different languages, and cartridge ribbons that permit changing ribbons in three seconds, with no inky fingers. Then, there were circular-dial type machines that, because of their shape, made it virtually impossible for the typist to see what he was typing. At last there was one, the Visible Writing Machine, that managed to solve that problem, and it actually looked very much like many of the manual models in operation today. It was the first to offer margin measurements, a bell sound when the margin was reached, and a carriage release. The Sears Roebuck catalog describes it immodestly as "not only perfection, perfectly perfected, but simplicity, simply simplified." Yet no matter how perfect or how simple, it cost more than most offices could afford, and most businessmen still relied on pens and pencils as their instruments of correspondence.

Fountain pens had already been invented, but many businessmen still preferred the straight pen, even with its built-in problems. It needed special penholders and it needed inkstands. In a sense, these were the office gadgets of that time. Simple penholders just held the pen point, but then imagination took over. One had a tip that could be adjusted to any angle convenient to the writer. Another came with a pen-ejecting holder so the writer could expose or not expose the point as he wished. The most unusual, the Myrograph, came with a special holder that attached to the hand in order to "prevent the finger movement and develop the muscular movement in writing. Keeps pen and hand in correct position. Prevents writer's cramp."

Inkstands not only became wells that held ink for the straight pen, but status symbols as well. Some were crafted with extraordinary designs, others were terribly minute, designed to fit into the vest pocket. There were those made in the shape of a guitar or piano, and the most complex doubled in function as receptacles for pens, points, pencils, stamps, paper clips, erasers, and other office minutiae.

There were novelty fountain pens for sale as well, including those that came in combination with lead pencils, erasers, knife and pencil sharpeners, and rulers. The status gadget of them all was the fountain pen that not only held ink, but also doubled as a cigar cutter.

A quick look at the Johnson Smith catalog of 1929 shows

that though the Depression was starting, the business people still maintained a sense of humor, which was reflected in the gadgets they bought. Two that took the country by storm were the 7-in-1 Combination-Pencil, which held a pencil, cigarette holder, put-and-take top, pen, compass, three dice, and as the *pièce de résistance*, a small microscopic picture that was magnified when seen through an opening in the pencil (the two most popular pictures were, not surprisingly, a naked female and the Lord's Prayer in type), and an automatic pencil gun, a curiosity shaped very much like a pistol. The pencil came out when the trigger was pressed and slid back into the muzzle or barrel when not in use. Knowing the economic situation in the late twenties, it was a good thing it was just a gadget and that only a pencil came out.

Our pens and pencils today aren't half that interesting. There are mechanical pencils that serve as calculators and automatically give the answer to addition, subtraction, multiplication, division, and square-root problems, pens that contain multiplication tables, and even those that light up in the dark for the businessman who might want to take notes at a meeting when a film is being shown. By and large, though, the humor has gone out of the simple gadgets.

Today's gadgets are more practical, created so that workers can effect maximum output for minimum input. Gone are the days when letters were opened by hand, postage affixed by hand, pencils sharpened by hand, papers destroyed by hand, tape dispensed by hand, even telephone calls dialed by hand. The hand, in fact, is in danger of becoming a useless appendage in the office. Gadgets are taking over.

Most secretaries agree that the Electric Letter Opener is a welcome addition to any office. It not only saves time, it eliminates annoying paper cuts that are a built-in by-product of opening letters by hand. These openers come in various weights and sizes. Some plug right into a socket, others operate by battery. The most interesting one by far is Invento's Cordless Triple Header, a combination letter opener, radio, and pencil sharpener.

How to Succeed in Business

Speaking of pencil sharpeners, the Apscomatic electric sharpener not only sharpens the pencil in seconds but, thanks to a dial that can be preset to one of a dozen selections ranging from dull to extra-sharp, the pencil can be, in a manner of speaking, sharpened to order. The gadget shuts off and a notification light automatically comes on as soon as the desired point is reached.

The electric stapler obviously saves a great deal of effort by the secretary who previously had to bang down on the old-fashioned staplers, which invariably got stuck just at the moment papers had to be collated to meet a deadline. However, a real innovation in the stapler family is the Stapleless Paper Stapler, which can fasten up to six bond sheets or fourteen onion skins with an invisible seam called a weld. The papers stay together until someone physically separates them, thus rendering obsolete another favorite little gadget, the staple remover.

Battery-operated tape dispensers are finally catching on and for good reason. Their major selling point is that since the press switch automatically cuts the tape off at the desired length, it eliminates the messiness of tape sticking to the fingers, which so often happens when it is torn off manually. Another tape dispenser that is beginning to sell is the Wristband Tote Tape. Everyone knows how impossible it is to wrap a package with one hand, hold the tape dispenser with another, and pull off the tape—with what remaining hand? Now it is possible by strapping the Wristband Tape-Toter to the wrist so that the hand that originally held the dispenser is freed to cut off the tape and affix it to the package.

Tape measures have come a long way from their infant days as glorified rulers. A popular combination is the All-in-One Tape, a spring-loaded ten-foot ruler that not only shows inches, feet, meters, and centimeters, but also has a retractable point that can become a compass or line scriber. The measure serves as a level and perfect square as well.

And for lefties, a special tape measure that pulls out to the right, so the numbers are read from right to left, and that rewinds counter-clockwise is available from The Left Hand store in New York City.

Electric paper shredders recently came to the attention of the public in connection with the Watergate affair, though they have been on the market for at least thirty years. In the last four years, sales have increased from 20 to 30 percent each year (which means there are between thirty and forty thousand a year being sold), spurred on by the enormous increase in the use of paper, the increased desire for secrecy, and the pressure from environmentalists against burning unwanted paper. The machines range from small hand-cranked models to a 416-pound number that is more than three feet high and three feet wide and can zip through 2500 pounds of paper an hour. This, the manufacturer adds, is the one that goes to the government.

At one point, the average-sized shredder, which normally fits over the wastebasket, became something of a status symbol among those executives who wanted to impress visitors with the fact that they dealt with important confidential material. One humorous innovation is a gadget that has a shredder on one side and a wastebasket on the other. The user can casually toss papers into the "shredder," impressing his visitor, only to retrieve them later so that he can go through them, making sure he hasn't tossed out something he should keep.

The telephone was originally a gadget. Today, phone gadgets are big business. Those that record messages are an agreeable alternative to depending on operators at a telephone-answering service who may or may not pick up the phone by a specific ring, if at all. There is one machine, not yet marketed nationally, that not only answers the phone and announces your pretaped message on the first ring, leaving thirty seconds for the caller to leave his message for you, but also permits you to change that prerecorded message at will by simply dialing a certain number and whistling a preset sound into the speaker. This signal then triggers the relay sequences that allow you to change the answering cassette.

The current best-selling device is the Phone Mate, which answers, delivers, and accepts messages and lets you monitor the incoming calls without letting the caller know you are in. It also has a very valuable partner, the Remote Mate, with which you can pick up your messages from the Phone Mate at any time from a telephone anywhere in the world. It plugs into the Phone Mate and is activated by your own specially coded pocket control. You just press the "on" button, dial your Phone Mate, and put it against the mouthpiece and get your messages. The Phone Mate then resets itself automatically for new message reception.

There are some machines that don't give or take messages but do record all incoming and outgoing conversations. Some use tape, others attach to the phone line and any standard cassette recorder. This, incidentally, is legal according to federal statutes as long as one party on the line is informed that the call is being recorded.

Let's look at what some other gadgets can do for you. The Tele-Caller provides for conference calls. You snap the device into a multiline telephone, dial one party, put him on hold, then dial a second party to connect to a three-way conference call. You can even hang up and let the other two talk, using your telephone as a bridge.

The Divert-A-Matic automatically transfers calls to any other phone number that you wish, which means you can finally receive important business calls whether you're at home, in someone else's office, or wherever you are having lunch.

The Time Limiter, attached to the phone, starts operating as soon as the receiver is lifted from the hook. After the first three minutes, and every three minutes thereafter, a warning beep is heard on the line, a great advantage to people concerned with the cost of long-distance calls.

A call waiting gadget can be inserted so that its beep lets you know that a second party is trying to reach you when you are on the line with someone else.

When you're away from your desk, you're not necessarily away from your phone. As long as you're within a half-mile radius, you can pick up your calls with a Dial-O-Matic cordless extension phone that you carry with you when you leave the office.

Beepers have been around for some time now and are especially useful to doctors and others needed on an emergency basis. With one model, when a call comes in to its owner's number, the beeper that he carries with him automatically sounds off so he knows to call back his answering service and pick up the message. With another, the Air Call, as long as he is within a forty-mile radius, the answering service operator can relay the message to him directly.

There are gadgets that amplify and gadgets that silence. One amplifier attaches directly to the part of the phone held to the ear and is of great benefit to the hard of hearing. It is portable and can be used on any phone. Another one is more complex. The phone is picked up in the normal manner and then placed directly on the cradle of the receiving and transmitting unit. The sound comes through a separate high-fidelity extension speaker. It is extremely valuable when you want everyone at a conference table to listen in and, in addition, it frees the hands so you can take notes on what is being said.

Regarding silencers, the Hush-a-Phone, which snaps on and off the mouthpiece, acoustically blocks out all surrounding noises and improves hearing at both ends of the line. Another device is a mouthpiece cap with a silencing button that allows you to talk privately to another person in the same room without covering the mouthpiece to prevent being overheard by the caller.

The Name Caller, which memorizes thirty-eight phone numbers that you use most frequently and automatically dials them for you, is this years "in" telephone gadget. It is an easily installed small unit that comes with an IBM electrographic marker. It not only eliminates direct dialing, thus doing away with dialing errors, but more important, if police, doctor, and fire department numbers are programmed, it can save the few seconds of time that might make the difference in an emergency.

Runner-up for "in" honors would have to be the automatic electric speaker phone. With this gadget, you never have to lift the receiver again. When you receive a call, you just press a button on the phone and the actual conversation takes place through a speaker. To make a call, the button is pressed once more, the number is dialed, and again, without lifting a thing, communication takes place via the speaker.

What next? Plenty. Bell Laboratories is working on a number of new concepts. Plagued by the problem of out-of-order pay phones as a result of vandalism (the cost runs into millions of dollars a year), they see a new "no-hands phone" in the future. There will be no receiver as we know it today. The caller will step inside a booth into a circle that is an electromagnetic field. He will insert a coin and press an "On" button that will activate a microphone and speaker. An automatic timer will disconnect the call when he leaves the booth.

Bell is also working on cassette telephones for sending messages to many predetermined numbers simultaneously; Dick Tracy-style, wristwatch phones; hand-free phones that are "dialed" with voice commands; home "sentinel" phones that detect fires, floods, and intruders; picture phones with copy printers that, at the touch of the button, produce a copy of what is being seen; and phones into which credit cards may be inserted to enable the caller to pay bills, order various kinds of merchandise, verify bank balances, etc.

A model of the portable phone of the future was introduced last year by Motorola after ten years of research and $7 million investment. Called the Dyna T-A-C, it weighs only twenty-eight ounces and looks very much like a military walkie-talkie. With it, a person can make or receive a telephone call from just about anywhere in the world. Numbers are called in the customary fashion on a push-button keyboard. Then the signal is transmitted over FM-radio frequencies to strategically located transmitter-receivers that feed the signal into the regular telephone network. The system works in reverse for incoming calls. The cost will be comparable to a car telephone, $60 to $100 a month, and will hopefully be on the market by 1976.

A section on office gadgets would be incomplete without a quick look at the latest in copiers and calculators. The new Xerox 400 Telecopier and the 3M Remotecopier make it possible to send and receive documents, photographs, graphics, and sketches over ordinary telephone lines. Portable at eighteen pounds each, they are being used by reporters, among others, to transmit copy, complete with handwritten notations and editing marks, to their newsrooms in distant cities.

Color office copiers are a reality at last, with Xerox and Hitachi the first to market machines capable of reproducing halftone colors with good fidelity. The Hitachi machine copies a color picture in ninety seconds and produces twelve duplicates automatically.

And finally, the Xerox 4000, the first to feature two-sided copying.

Calculators make it possible to quickly and correctly solve just about any mathematical problem an individual is faced with. They come in all prices and sizes, with the Sinclair being the tiniest in existence. It weighs two and one-half ounces and is only one-eighth inch thick.

The person who pays the bills in a small company that can't afford a bookkeeper might use a Dial-A-Balance, a small adding-machine computer that fits right into the checkbook. One dial is set to the present balance, another to the amount deposited or deducted, and the new balance automatically appears in a special window.

A firm that deals with great amounts of cash and is concerned with the authenticity of the bills will be grateful for the Astra-Lite or Money Monitor, under which counterfeit currency can be automatically determined.

Financial officers requiring banking services after hours may choose a bank that utilizes the Ultra/Matic 24, a mechanical bank teller that can handle 80 percent of all banking needs twenty-four hours a day. A special card and code number is inserted into the machine that permits the card's owner to make deposits, pay bills, receive cash, and even transfer funds from one account to another. If an improper code number is entered, the machine provides three chances to correct the mistake. If there's no correction, the machine captures the card and keeps it.

A secretary with cold feet might welcome one of the various electric heat pads that fits conveniently under the desk. She might also appreciate the new self-correcting typewriter ribbon, the bottom half of which contains white ink. When a mistake is made, she just has to back-space, shift the ribbon selector, and type over the error.

An executive with branch offices around the world will be grateful for the Data-Timer, which shows what the time is locally in whatever country he may want to call.

And a boss with a need for inspiration first thing in the morning might smile at a digital musical desk calendar that plays "Oh, What a Beautiful Morning" as he changes the date each day.

Finally, the ultimate in office gadgets—the Mini-Cart. This is a veritable office factotum that can deliver coffee as well as pick up and deliver mail, memos, etc. It travels a route of magnetized wires buried in the floor, moving silently and automatically from one location to another. It can call for an elevator by itself, yields the right-of-way to human beings crossing its path, and even returns to its battery charger when on-board signals warn it that it is low on juice. Watch out. The next thing you know, it will be after your job!

GETTING THERE IS HALF THE FUN

Travel Gadgets

Travel is one of the country's leading industries, and taking to the road, oceans, or skies, be it for business or pleasure, is very much part and parcel of the American way of life.

Gadgets for the automobile never cease to intrigue. Since some commuters spend as much of their time during the week with their cars as they do with their children, it is no wonder that they take an active interest in any gadget that will make their ride safer, more interesting, and possibly more enjoyable.

Car thefts are unfortunately on the uprise (over 1 million cars will be stolen this year), and it has gotten to the point where some sort of burglar alarm or safety device is almost as important to the driver as the steering wheel. Some of the simple ones lock the steering wheel to the brake, others lock the hood, and still others send out an alarm, but the sad fact is that they are only partially effective.

Studies show that 76 percent of all car thefts are the result of keys being left in the car. One of the newest safety gadgets to counteract this is the Auto-Guardian, a solid-state subminiaturized electronic computer system designed to prevent theft even when the thief is in possession of the proper key. This is how it works: The computer keyboard, mounted to the dashboard, is programmed for a four-digit number combination that is known only to the driver. Even though the thief may have the key, if he doesn't know the code, he can't start the motor. The Auto-Guardian costs more than other protective devices, but owners say the expense is well worth the security of knowing the car will be found where it was left.

The one thing all drivers want to avoid is an accident. In terms of gadgets, this is being approached in two ways: by warning systems in the car itself and by special diagnostic devices to help the auto mechanic pinpoint problems before, not after, they arise. One interesting driver's warning system is a unit placed on the rear window ledge and controlled by a remote-control box plugged into the cigarette-lighter socket. By flipping the appropriate switch the driver can warn motorists behind him to DIM LIGHTS, STOP TAILGATING or, in the case of emergency, SOS.

The Japanese have an unusual device to cut down on accidents when parking. A tape recorder is mounted in the car's trunk with a speaker on the rear fender. When the driver shifts into reverse, the tape automatically begins to roll and a voice is heard saying, "Please be careful. This car is backing up."

If all goes well, rear-end collisions will be greatly reduced by RCA's experimental auto radar, which will be unaffected by rain, fog, dust, or smog. A radar unit will be mounted on the front of the vehicle and a reflector on the rear. Signals will bounce from one car to another, enabling compact calculators to compute the distance between them and relay this information to the driver via a meter on the dashboard. If the distance should grow too short, a warning light will flash or a buzzer will sound, and hopefully, accidents will be averted.

In this same vein, a patent was recently issued for a collision sensor called the Barbi (Baseband Radar Bag Initiator). It detects signals at first and second distances between the vehicle and the other object, the speed is taken into account, and when the car is ten feet from the other vehicle, it makes a decision whether or not to release an airbag or other safety device.

Since the majority of traffic accidents are caused by drunken drivers, the United States Department of Transportation's Office of Alcohol Countermeasures is currently trying out a portable roadside testing device that will show the police instantly how intoxicated a person is and whether further testing on a Breathalyser at headquarters would be appropriate. Alert (Alcohol Level Evaluation Road Tester) is a portable two-pound self-contained electronic unit with green, red, and yellow buttons that light up or don't light up (depending on the amount of alcohol ingested) when the suspected drunken driver blows into a plastic tube. It is hoped that the officer will then be able to convince the driver he has had too much to drink and to slow down *before* the accident.

Too many accidents are caused not by careless driving but by malfunctions in the car that should have been caught by a mechanic. It has been estimated that from $8 billion to $10 billion a year is wasted on work not done or on repairs that were not needed but done because the mechanic didn't know what was really wrong. At last, sophisticated diagnostic equipment, similar to the electrocardiograph in medicine, is coming into increasing use to take the guesswork out of what is wrong with cars.

Already a red warning light on the dashboard indicating low brake fluid is standard equipment on many 1973 models, and the General Motors Corporation has approved, as a dealer-installed option, a sensory warning device that shows insufficiency in oil supply and brake fluid and also indicates brake-lining wear.

And Volkswagen dealers are offering owners diagnostic printouts for a variety of vehicle functions tested by connecting a permanently wired test harness and central socket to a computerized diagnostic device in a Volkswagen-dealer's service shop.

Volkswagen's so-called "offboard" approach has been modified in the experimental test system Toyota calls the Electro Sensor Panel. Sensors throughout the vehicle are connected to a special panel in the dashboard, and an electronic diagnostic unit at the dealership pinpoints the exact trouble area for proper servicing. More than two dozen sensors are used in the Toyota tests. If the oil indicator is lit, for example, the dealer's diagnostic device specifies whether the trouble lies in the transmission, the engine, or the power-steering pump.

General Motors has been intensively exploring a total "on board" approach to diagnosis. Under the G.M. system, car owners would be able to ascertain certain trouble areas through sensors and possibly a computerized data panel without the need to "plug in" at a G.M. dealership. It will be a few years before all these systems are perfected, but it is a welcome step in the right direction.

Not only does the driver have to cope with the possibility of careless driving or car malfunctions, he must also face the challenge of the elements. Ice and snow on the rear window and windshield are definitely not the driver's best friend, but when they appear, the various defrosting devices invariably are. The most popular is the car defroster gun that, powered by the 12-volt battery in the car lighter socket, clears the front and back windows instantaneously. Another gadget that accomplishes much the same thing is the Rear-Window Defroster, an electronic printed circuit wired to the ignition switch that fits directly on the back window. It, too, operates on a 12-volt battery and will melt snow or ice at the flick of the switch. A third alternative is an electric windshield scraper.

Plugged into the cigarette lighter, it heats up in ten seconds. The heated-rod part applies the heat, and the squeegee-rod part wipes the window clean.

One way of avoiding ice and snow on the windshield, of course, is to have a gadget that will prevent it from accumulating in the first place. There is one called the Magnetic Windshield Protector, and it is exactly what the name implies. It is a heavy opaque plastic cover that is put over the windshield and held in place by magnets at all four corners. In the morning, all that needs be done is to remove it, and the windows are perfectly clean.

Speaking of the elements, another little gadget to keep in mind, especially when you're stuck in heavy snow and can't move, is a thirteen-ounce lightweight steel "traction aid" that provides the necessary traction when it is slipped under the rear tires.

Let's look at some of the gadgets that make traveling by car much more pleasant than it would be without them. If the climate is too hot, the driver can cool off with a car seat built with a powerful electric blower that plugs into the cigarette lighter and circulates air all around him. If it's too cold, he can warm up on a "hot seat" that automatically radiates heat when touched. The colder the temperature, the hotter it becomes.

A fun gadget to use at the toll booth is the toll gun. Prefilled with change, a pull of the trigger sends the proper amount of money soaring into the receptacle. Another aid for those who pay tolls often is the Quick Change Dispenser, a spring-action disc holding two dollars in nickels, dimes, and quarters that fits into a self-stick mounting base on the dashboard.

A favorite with drivers who savor that extra cup of coffee when driving to work is the beverage caddy, which attaches to the underside of the car's dashboard. This plastic unit, which holds a twelve-ounce can or cup, adjusts so that it is always level, is guaranteed not to tip, and is spill-proof.

To help keep the car clean, a popular item is the lightweight electric Auto-Va, which sucks up accumulated dirt with a minimum of effort.

A Radar Sentry, through long-range transistorized antennae, alerts drivers far in advance to the fact that they are approaching radar zones and should slow down.

And for those who prefer having things done for them, there is the Genie, a remote-control gadget that automatically opens the garage door, turns on the light, and then closes it once the car is in place.

A patent with far-reaching consequences was recently taken out for a device with which a person can keep tabs on his electrocardiogram while driving. He grips electrodes that are inlaid on each side of the steering wheel rim and that are also connected to an instrument in the center of the wheel. If the driver's heart rate falls outside his prerecorded norm, he gets a warning. The instrument is not expected to render clinical judgment, but it warns the subject that it is time to see his doctor. To date, it is still being evaluated.

One of the gadgets that has attracted some press interest is the so-called "hot seat" in which the weight of a passenger entering a taxicab trips a recording device that tallies the number of riders who enter the cab each day for the fleet owners. This is a management check to see if drivers are cheating them by offering riders "compromise fares" that the drivers don't have to pay commissions on. As with most gadgets, there are ways to get around them. Ingenious cabdrivers have learned how to deactivate them; others simply invite a prospective passenger to sit in the ungadgeted front seat.

Which cars have the most interesting gadgets today? And what gadgets will be part and parcel of the cars of tomorrow?

I doubt that any cars offer more than the famous His-Hers Rolls-Royce twosome, which, before inflation, was retailing at $250,000 the pair. Some of the interior gadgets the owners can point to with pride include stock tickers, tinted reading lamps, air conditioners, telephones, liquor cabinets of matched deep walnut stocked with the finest of Waterford crystal and vintage wines and liquors, quadraphonic music systems, automatic dictating devices, electric dividers, and especially designed electric sun roofs in the rear compartments.

The Silver Eagle is one of the few racers powered solely by electricity. With a top speed of 152.598 miles per hour, it uses the same battery system as the Apollo moon buggy. It contains 180 silver-zinc cells mounted in the rear, enough power to keep a car humming at expressway speeds for several hours. The gadget is the battery. The cost for each one, four thousand dollars.

And the cars of tomorrow. There are two ways of looking at them—practically and impractically. Practically, the British government and its automotive industry is looking into an auto bumper designed to protect pedestrians (40 percent of all highway casualties), a headlight that tilts automatically to keep the beam level up and down hills, a system that projects driving information onto the windshield (speed, oil pressure, etc.) so that the driver doesn't have to take his eyes off the road to check the instrument panel, and an audio information system that begins broadcasting warnings such as accidents, fog, a washed-out bridge, or detours, to drivers whenever an automobile passes over transmitters embedded in the highway.

At the latest International Automobile Show, a group of engineering students at Brooklyn's Pratt Institute submitted their design for the urban car of tomorrow, a modified suburban station wagon about the size of a Volkswagen Squareback, 160 inches by 60 inches. To improve the driver's rear-view vision, a TV camera with a 150-degree sweep that automatically adapts to lighting conditions was mounted on the rear of the car. To protect the driver, even against his will, there was a Drunkometer, a brake-pedal time reaction test that the driver would have to pass or else the car would automatically lock itself for fifteen minutes before the driver could try again. Instrument panels would enable the driver to monitor all of the car's functions at the same time. Finally, it would have vinyl bands with steel tubings to protect passengers in the event of a side collision.

89 • Getting There Is Half the Fun

Psychologists, both amateur and professional, look upon the selection of automobiles as the expression of unconscious desires on the part of the purchaser. A convertible symbolizes a mistress. Big connotes greatness. Noise reinforces masculinity. And advertising agencies have gone a long way to reinforce this. Cyril Kornbluth, a noted science fiction writer, took these things into account when asked to describe his car of the future. He replied, "It will be a big car. 40 feet long. Huge tail fins. Enormous fins. Incredible noise." He then added, laughingly, "And even though it revs up . . . to . . . 22 mph, no one will care. It will still make you feel like you're doing something big . . . and at the same time, you won't be able to hurt anyone."

And when it comes to sheer impracticality, nobody has been able to equal the Whimsymobile, the futuristic new motoring carriage created by noted British cartoonist Rowland Emmett for New York commuters tired of contending with parking tickets, high gasoline bills, and noxious fumes. Called the "Vintage Car of the Future," it operates in place all the time and runs on boiled after-shave lotion that produces an exhaust so fragrant it attracts clouds of butterflies. The engine is powered by a bronze cherub that delivers a blast of oxygen into a revolutionary turbine with coffee-spoon blades. For long-haul trips and hungry youngsters, there is a built-in barbecue. And to make driving safer during rush hours, there is a crystal ball to tell what the motorist ahead is going to do.

It should only happen!

The travel industry has experienced an incredible growth over the past ten years. Travel gadgets have been created that not only weigh little and take up almost no space at all but in addition provide the wanderer with almost all the comforts of home. For instance, it is possible to pack a clock, hair dryer, travel iron, and dry shaver, and together they will weigh less than four pounds. The trouble with trying to get away from it all these days is that it all comes along!

An initial problem facing travelers is who will handle the tasks that normally require their personal attention. Who will water the plants, feed the dog, or see to it that the goldfish will be alive when they return? Gadgets will.

There are a number of devices that will water the plants for you for up to thirty days. One is a flower-pot holder that includes an automatic watering device. When the pot is placed in the center of a white plastic square and the opening is filled with water, the plant will absorb as much of it as it needs for nourishment. Freddie the Frog is another device that waters your plant automatically; once he's filled with water and his stem is inserted in the soil, the plant will absorb just as much water as it needs. A third favorite is the Floramatic Plant Quencher, which holds enough water to supply the plant for at least three weeks.

Your dog or cat won't got thirsty as long as you own one of the many pet waterers or drinking fountains that automatically refill whenever the animal drinks from it, and you can see to it that the fish are taken care of during your absence by installing a lightweight adjustable fish feeder on the inside of the aquarium. It holds up to a month's supply of flake or granular food and dispenses it automatically to the fish each day.

Until a few years ago, the one gadget that was indispensable to any traveler planning to use an appliance like an electric razor or traveling iron when he was abroad was the convertor-adaptor, which transforms the foreign electrical power of 220 volts to the 120-volt power by which most American gadgets operate. However, in the past few years manufacturers are catering specifically to the travel trade by making items that are not only lightweight and portable, but that operate under both currents, thus rendering the transformer almost obsolete.

A situation familiar to every traveler at some point or another is what to do with clothes that become wrinkled while in the suitcase. At last, there are a multiplicity of solutions. Norelco has combined the small travel iron with a lightweight presser. For fabrics that can be steam-pressed, Invento has a model that does just that. Oster offers a little steam iron with Teflon-coated plates; a tiny sewing case is also included. And for replacing lost creases in pants, there is the Miniature Pants Presser from Westinghouse.

Perhaps the favorites of the new travel gadgets are the portable hair dryers and stylers that make it possible for both men and women to use the time they previously would have spent in the barber shop and beauty parlor to do more interesting things. The Worldwide Electric Barber for men offers any service a man could receive from a professional. It consists of an electric razor along with two extra heads that do everything a barber's clippers do; a clipper to sculpt the beard, dress a mustache and sideburns, and shave the neck; and another clipper that transforms the machine into a hair cutter with a fine, medium, or long grind.

For women, the twelve-ounce Braun International Hair Styler is a godsend. It comes with a comb-and-brush attachment for styling the hair while it is drying. It is often used in conjunction with the World Wide Hair Stylist electric comb that lets the woman set and shape her hair quickly without having to resort to unflattering, time-consuming rollers.

Some interesting timepieces have been created for the traveler. The most unique is the world's smallest alarm clock (one and one-half inches by one inch), an electronic model with a buzzer that works on a battery. Some of the clocks show the exact time both locally and at home. Ardate, a Swiss company, puts out a wristwatch that does the same thing. The World Travel Alarm not only shows local and home time but, in addition, the time in every major city around the world. And for those addicted to the radio/alarm clock, General Electric has a one-pound model that will make you feel like you never left home.

Finally, a quick potpourri of other travel gadgets on the market—some familiar, others not so:

For those who carry a cane, a seven-ounce foldable self-opening model that fits into the pocket or purse.

For the person who can't wait for room service in the

morning, the Worldwide immersion heater that makes cupsful of hot water in seconds to which you can add instant coffee or tea.

For the man or woman who doesn't like to be without it, a folding shoe horn.

To keep the wig in shape, a stand made of wires that folds flat for packing and forms into the shape of a head when assembled.

To liven up a salad, a two-and-one-half-inch pocket pepper mill.

Lastly, the Baggage Master, a "do-it-yourself redcap" with a set of strong, lightweight wheels to slip under the luggage and pull if help can't be found.

Without a doubt, gadgets can make a trip a more delightful experience. In fact, gadgets can be a trip in themselves!

9

A SPORT IS A SPORT IS A SPORT

Gadgets for Fun and Games

Do you remember back ten years or so when adults would argue with each other whether or not the box that sat on their desk with red lights flashing on and off at random should be called the "something" box (because doing "nothing" was actually doing something) rather than the Nothing Box, which is what it was originally named? Or can you remember back even further when people would spend money on silly items like cocktail shakers that would play "How Dry I Am" when they were wound up? Now, there's a new "nothing" gadget on the market, a game of sorts called Cascade. It consists of a little box, three rubber "drums," and a target. The player simply presses a button that sends the balls on their bouncing way to the target. The player has absolutely no control over the thing—the balls just go—and then he pushes a button that mechanically sends the balls back from the target, through a plastic tube, to the little box. It may sound feeble, but it's practically addictive.

These are but three examples of gadgets that exist for one reason only: they are fun.

Some reflect the interest that the younger generation, especially, has taken in the world of psychedelics. These gadgets include simple flasher units that plug into light bulbs or lamps and cause the bulb to give off eerie, rapid, psychedelic effects, "psych lights" that flash weird, bizarre color designs around the room, and electric sound to light converters that change audio sounds from the stereo or radio into light signals that create a "musical light" of psychedelic effects.

Popular among executives are the various "decision makers," for those who need all the help they can get. Some consist of little balls operated by magnets that, when the operating button is pushed, either fall in the Yes or No hole. Others swing from a fulcrum before they finally stop over Yes, No, or Maybe to guide you as you make your decision. Chances are any one of these is as right as any executive himself is after a three-martini luncheon.

Probably the most inventive of the fun gadgets is the Sonus Switch, which operates much like a remote control. All the owner has to do is clap his hands or snap his fingers and anything using up to 750 watts of electricity that is attached to it is turned on automatically via the high-frequency sound waves created by the noise. One bachelor had great fun fitting this to his own specific needs. He attached it to his tape recorder on which he had pretaped a message. Upon coming home with a girl friend, he would clap his hands and go to the bar. Suddenly, from out of nowhere, she would hear his voice saying, "I hope you've enjoyed yourself as much as I have. Please, make yourself at home while I pour us a nightcap. The evening is just beginning." A perfect example of a fun gadget being taken seriously—and used for fun.

It reminds me of a gadget I saw in the film *The World of*

Henry Orient. In one particular scene Henry, a rather degenerate pianist played by the inimitable Peter Sellers, sat down at the piano to play for a woman visitor. The music built up dramatically to a pounding crescendo, whereupon Henry flipped a switch and the music continued without a break as he rose to join the woman on the sofa.

Martha Mitchell may symbolize many things to many people, but no matter what their personal feelings are, collectors of fun gadgets will always be grateful to her for being the inspiration for a best-selling conversation piece—the oversize telephone. Eighteen inches long, four inches wide, and made out of black foam rubber, the real mouthpiece can be cradled in it so you can actually hold this incredulity and carry on a normal conversation.

Another fun gadget being found in offices is the desk lamp whose base is a miniature parking meter. A nickel sets the meter for an hour, a dime for two hours. At the end of the allotted time, an arrow shoots up saying "Time's Up" or "Violation," and the lamp automatically turns off. One executive I know values it as more than a conversation piece. When he has a meeting in his office that he doesn't want to run over an hour, he puts in a nickel and sets the clock in an inconspicuous place. When the light goes off, he knows the hour is up, and he can bring the meeting to a close without making his guests uncomfortable by referring to his watch.

A hobby that is attracting more and more aficionados is hunting for buried treasure. Not too long ago, *The New York Times* carried a story about a man who unearthed a two-hundred-year-old, twenty-pound fragment of the most famous piece of sculpture in American history, a large equestrian statue of George III, the last king of the American colonies. Louis Miller, the man who found it, told reporters he made the discovery with the aid of a metal treasure locator. Some of these work with electric coils, but most of the treasure detectors operate via transistors. Supposedly, they send electronic beams through earth, concrete, rock, wood, or water, and signals are sent up whenever anything buried, be it coins, jewelry, silver, gold, or even statuary is come upon. Not everyone is as lucky as Mr. Miller, but the popularity of these Snooper-Tronics proves that a lot of people are having a lot of fun trying.

As children, we were introduced to gadgets through one of our favorite recreations, the comic book. Who can forget Dick Tracy keeping in touch with headquarters via his two-way wrist radio? When he was doing this as far back as 1946, it was considered beyond the realm of reality and was often put down as science fiction. Today, of course, what was then science-fictional has become science fact. It was also through Dick Tracy that we learned about the "telegard," then, as now, used to monitor possible criminal activities, only today we call it closed-circuit TV. Tracy, too, familiarized us with the two-way wrist TV; the atom light, one thousand times brighter than sunlight, used to thwart robberies; and the voice-print machine, all of which are now, of course, reality.

Today's youngsters are learning about gadgets from DareDevil (who, using a billy-club cane especially designed with hidden microphones, transistors, and tape recorders, can overhear conversations miles away and detect potential criminals before they have a chance to perform their vicious acts), Spider Man (who captures law breakers with his unique spider-web shooter), and Iron Man (whose arm, legs, and body are covered with weightless transistorized magnets that, when put into action, give him the strength of Samson).

It is impossible to refer to any comic books or cartoon strips, of course, without paying homage to the dean of fantastic (in every sense of the word) visual gadgets—the late Rube Goldberg. Master of the absurd, he created, in drawings, incredibly complicated devices that, once understood and mastered, would lead to the inevitable—nowhere.

One game that is enjoyed by just about everybody is the game of cards, so it comes as no surprise that there are a number of gadgets designed to make card playing as enjoyable as possible.

Do you have difficulty shuffling the deck? Buy a push-button card shuffler. The cards are divided on either side of

97 • A Sport Is a Sport Is a Sport

the gadget, a switch is flipped, and presto, they are mixed into a single deck. Tired of dealing? Use a cordless electric card dealer that deals automatically for up to four players. Do your hands get tired holding the cards while you're awaiting your turn at gin rummy? Get a special gin-rummy card holder that will do it for you. In fact, now you can even play cards without the cards. Just get an electric poker machine; you press a bar and the symbols and number of five cards appear in the showcase. If you want to draw a card, you push another button. And you can bluff just as well with the machine as you can with real cards.

If bingo is your pleasure, look into the Electric Bingo Blower. The numbered white Ping-Pong balls are mixed thoroughly in the blower, and the new ball pops up automatically the moment the previous ball is removed and the number called, thereby speeding things up so you can play more games in less time.

Electronics have also caught up with the dice via portable electric dicers that operate via push button. Users say this eliminates the confusion of wildly thrown or lost dice, and again, saves time.

Not everyone considers doing crossword puzzles a game. To many, it is more like work. But whatever they call it, they agree that one of their favorite gadgets is the crossword-puzzle board that magnetically holds pens and pencils, erasers, a small dictionary, pencil sharpener, magnifying glass, and light, all the tools craved by the crossword-puzzle addict.

Sports sport their own share of gadgets for those who are active as well as for those who prefer to enjoy their sports from the sidelines. Little Leaguers, or their fathers, can practice batting on their own with an inexpensive automatic baseball-pitching machine, a mechanical robot pitcher, adjustable as to speed and height, that serves up balls automatically.

The tennis player has his counterpart in the Tennis Teacher, which can be set up in backyards or driveways and tosses the tennis balls ten times a minute as far as forty-five feet at an average of forty to fifty miles per hour.

Skiers are very enthusiastic about a new Skidometer that

registers speed downhill from eighteen to fifty miles per hour, stopping automatically at the highest speed reached. It's perfectly accurate, shatter-proof, totally unaffected by snow or water, and invaluable to racers trying to get in shape.

Gadgets for the fisherman are many and varied. Several, including the fishing thermometer, which rises as it is exposed to more and more fish, help the angler in his quest for "the perfect spot." The Lowrance Fish-Lo-K-Tor serves the same purpose but is a more ambitious effort. It is a portable device that accurately reports the size and depth of the school of fish so you know immediately whether you're on top of a minnow or on top of a shark. And, fortunately for industrial societies, it works in muddy water as well as it does in fresh or salt water.

Once a fish is spotted, however, it is not automatically caught. To make it easier to land the catch, many fishermen use the Insta-net, a full-sized fishing net on a flexible frame that, after use, folds with a twist into a six-and-one-half-inch leather pouch to be worn on the sportsman's belt.

Once caught, the next step is preparation of the fish, and, of course, it comes as no surprise to learn that in this endeavor, too, there are gadgets to help. Fish scalers are too numerous and prosaic to mention. One rather unique item, however, is a fish skinner that pinches off the skin between the blade and the roller. You turn the handle and—presto—the fish is skinned.

And for those who prefer to remove just the scales and leave the skin on, there is a stain-resistant hardwood fish-cleaning board available with a toothed spring clamp that holds the fish securely by the head, making scaling a cinch.

The hunter, too, has inspired his share of gadgets. As with many other sports, many of these are designed to keep the enthusiast in shape during the off-season. One such aid for the bird shooter is a target launcher that, using 22-caliber blanks, can send empty beverage cans flying as far as forty feet up- and forty yards down-range. The Spot-Shot is for gentler practice. It is a battery-powered light, in the shape of a shotgun shell that is actually loaded into the barrel. When the gun is "fired," a spot of light appears where the gun is aimed, and the size of the spot lets you determine how accurate you were.

Golfers wax eloquent when describing the automatic score keeper which, worn like a wristwatch, clicks off every stroke automatically up to ninety-nine strokes. Some say they wax eloquent because it doesn't go above ninety-nine. Another favorite among golfers—some say it boosts their driving distance 20 percent—is the electric golf-ball heater, which warms golf balls up to six hours before use. These golf balls, they insist, take off hot, literally and figuratively.

One problem that often plagues players once the ball has taken off is just where it has taken off to. The problem has been solved not by the caddy but by the Bleeper Electronic golf ball. Via a small radio transmitter embedded into the ball, signals are sent to a receiver the golfer carries in his pocket. The closer he gets to the ball, the louder the bleep, thus eliminating a great deal of guesswork and wasted time.

Any golfer who takes his game seriously (is there any other kind?) will appreciate a practice club called the Swing Trainer, which clicks to tell the golfer whether his timing is early, late, or right on target. It extends for indoor or outdoor use, collapses for travel, and the manufacturer claims it increases driving distance ten to thirty yards and can possibly cut up to fifteen strokes off the final tally.

Also gaining in popularity is the all-in-one golf club. With the twist of a dial, the head converts from a driver to a putter to a three, five, six, or seven iron. And a left-handed model is available for those who need it.

Another gadget golfers seem to enjoy is the divot fork and groove cleaner, which levels on-the-green divots in seconds, cleans shoe cleats and balls, and scrapes the grooves of the irons clean. There's also an adjustable, reusable two-position tee in the shape of a badminton shuttlecock that holds the golf ball much more securely than the ordinary tee.

For the golfer who will go on the links no matter how cold the weather and for observers of many cold-weather sports, there are many choices of foot warmers, pocket hand warmers, and electric toe-and-finger warmers available. New, however, is the Portable Air Warmer, which is, in effect, a plastic warming helmet strapped over the head and under the arms, operating by means of a tiny fan, special lamp, and original heat transmitter, that lets you breathe in cold air but automatically heats it to the temperature you select.

On the other hand, for those who go eighteen holes no matter how hot it is, there are pocket-sized portable electric fans to cool them off and a new Hot Weather Hat, a hatband attached by rigid poles to a sun shield, which protects the wearer from the sun while allowing a free flow of air between the hat and the top of the head.

For those who wouldn't mind playing in the rain, if only they could hold the umbrella and putt at the same time, there is finally a solution—a Portabrella umbrella holder that is strapped like a harness across the shoulder and around the waist and holds the umbrella stationary without any help from the hands.

If mosquitoes are a problem on the back nine, you can finally eliminate them with an Electronic Repeller that fits into the palm of the hand and drives away mosquitoes with high-frequency sonic vibrations at the push of a button.

And finally, when you get tired or are waiting for another foursome to tee off, how wonderful to unfold a two-pound padded vinyl seat cane, which opens onto a tripod base. If you have to wait your turn, there's no reason you shouldn't wait it out in comfort. Gadgets, we salute you.

MOTHER, PLEASE, I'D RATHER DO IT MYSELF
The Do-It-Yourself Gadgets

Gadgets reflect the fantasies, the realities, and the directions of the society in which they exist. Obviously, 150 years ago there would have been no place—no need—for, let's say, gadgets that defrost the refrigerator electrically in minutes before frozen food has a chance to thaw, or defrost alerts that let you know when freezer temperatures are changing and thawing is occurring. There were no refrigerators. By the same token, there would have been little interest in special "drip-dry" hangers before the various polyester fabrics were created.

A look at gadgets on the market today indicates a trend that is attracting more and more people, elderly and middle-aged as well as young, the "do-it-yourself" phenomenon.

There seem to be three reasons for this. One is simple economics. Thanks to the innovation and perfection of certain gadgets, it is much less expensive to do-it-yourself than it is to go to professional sources. For example, why spend hours in a beauty parlor when the electric hot rollers, wave setters, and portable hair dryers make it just as convenient, simple, professional, and less expensive to do at home? Why should you have to go to a shoemaker to enlarge a pair of shoes and return to pick them up at another time when for half the price and probably a fifth of the time, you can buy a shoe stretcher and use it as often as you want—at your leisure? Why pay more for expensive prepackaged herbs and spices when you can grow and grind your own at home?

Food prices have soared so high, so quickly that even the government got into the gadget act; in 1973, Mrs. Virginia H. Knauer, then President Nixon's Advisor on Consumer Affairs, unveiled the Tella-Cost, which helps the consumer keep track of just how much he is actually spending for each serving of food. The Tella-Cost is, in essence, a four-by-eight-inch slide rule that describes the various cuts of meat and grades of fruits and vegetables, lets you calculate the price according to weight through a viewing window, and indicates how many servings, for instance, one pound of hamburger will yield and the exact cost of each serving.

The second reason for the do-it-yourself craze has to do with the factor of time itself. People have more time on their hands than ever before. Some major companies are experimenting with the four-day work week, providing an extra day of leisure for many. High school and college curricula have changed a great deal, and fewer hours are being spent in the classroom than in previous years. Microwave ovens and prepackaged and frozen foods have contributed much to minimizing the time spent preparing food for the family. As a matter of fact, gadgets might even take a certain share of the credit for the extra time available. And now we will find gadgets taking up the slack and seeing to it that we have something to do in the time being made available.

The third and perhaps most driving consideration behind the do-it-yourself direction is the need many people have to feel they are part of their environment in as many areas of life as possible. They desire a feeling of accomplishment and are more than willing to invest the time and energy needed to create things from the beginning. They are up to their necks in artificiality and superficiality. They want to "do their own thing." To many, especially in terms of nutrition, this means going back to nature, using only organic or home-grown products, no artificial additives, no preservatives, just plain, simple home cooking, starting from scratch, just as their forefathers did, with a little help, of course, from their friends, such as refrigerators, freezers, and the many gadgets that have been developed to eliminate some of the harder labor.

It is said that bread is the staff of life, and it seems that making one's own is a starting point for many do-it-yourselfers. Many stores are now stocking home flour mills, which, they claim, save precious nutrients. Now it is possible to grind whole wheat flour, corn, oatmeal, rice, chestnuts, lentils, peanuts, cottonseed, or barley—a pound a minute—by hand, just as it was done in the 1800s. Those who want the advantages of natural nutrients but don't have the time or energy needed to crank a home flour mill can do it the modern way with an electric grain mill. Plug it in, put in the grain or seed to be ground, push the button, and in just seconds—flour.

Once you have the homemade flour, it's time to convert it to bread or rolls. Thanks to gadgets again, should you not wish to, you no longer have to manually knead your own. Buy one of the many automatic bread makers, mix the ingredients, put them in the pan, turn the handle, and the rotating mechanism inside kneads the dough for you. There's nothing left to do but put it in the oven and enjoy it when it comes out.

Yogurt is considered an exceptionally healthy food, so it is not surprising to discover that yogurt makers are very much in demand. Owners maintain that by making their own, they save up to 70 percent of what they would have to pay at the market.

Calories and cholesterol levels are of great concern to so many that it is not unusual to find people buying gadgets that let them make their own fresh cream using margarine and skimmed milk instead of butter and cream and their own mayonnaise and cake fillings according to their special dietary needs.

For those not on diets, there are a variety of pasta and spaetzle makers to choose from so owners can make their own spaghetti, vermicelli, ravioli, macaroni, noodles, and egg and liver dumplings and are no longer tempted to spend their money at restaurants that advertise "homemade pasta."

Let's not overlook the people who are into the do-it-yourself thing not for the financial savings or the nutritional value, but for the sheer fun of it. These are the people who are buying gadgets to make their own cheese and wine. And for those so inclined, there are electric bottle cutters with which, by pressing a button, glasses, candle holders, and other decorative pieces can be fashioned from just plain ordinary bottles.

Is there anyone who doesn't want to live in a cleaner, fresher environment? I believe every individual has to take some responsibility to do whatever little he can to contribute to it and, fortunately, there are special gadgets available that can help make day-to-day living just a little bit more clean and enjoyable.

In every major city, there are groups being formed to prevent the abundance of dog litter on the streets. There are a number of aids on the market for the pet owner, the most popular being the Poop-Scoop that makes it very simple to pick up the debris without spilling it and deposit it in a convenient receptacle. Mrs. Iris Sutton of Manhattan has recently designed a Doggie John, a lightweight indoor toilet for pets complete with a sketch of a fire hydrant. For those who live in the suburbs, there are small mini-septic tanks, which are installed in the ground, liquify the animal waste, and are odorless and insect-free. For cats, there are special potties that are nonbreakable, prevent litter spills, and are also completely odorless. It is interesting to note, incidentally, that this is not a problem that has developed overnight. The first "device for preventing dog nuisance" was patented in 1909. It was a device that was to be placed at convenient locations and would administer an electric shock when an animal attempted to "commit such a nuisance"—an early attempt at behavior modification, so it would seem.

What to do with trash concerns a great number of homeowners. Garbage disposals help to a degree, but not enough. But finally, thanks to a gadget called the Minimizer, it is possible to compress one week's trash accumulation from an average family of four, including cans, bottles, and kitchen garbage, into one odor-free single-lock folded bag—surely a step in the right direction.

The gardener concerned with leaf burning can eliminate the problem altogether with a shredder-bagger. With the help of this gadget, dry leaves and twigs are shredded and compressed to a tenth of their volume. The result is bagged, in the same operation, into neatly packaged compost makings that can then be returned to the soil.

Another problem plaguing big cities is noise, and the control of it is rapidly becoming an issue in many areas. It is now possible for the concerned citizen to buy a pocket twelve-ounce portable noise-pollution meter and keep his own record of noise levels wherever he is. As the meter also indicates when the sound levels exceed the limits specified by law, he can then, with proof, report back to his local representative and demand that something be done about it.

Unclean water plagues an even greater percentage of our population, which explains the popularity of two relatively inexpensive gadgets: one, a water filter that connects directly to the faucet and clears out the impurities before the water reaches the glass; the other, a special plastic electric appliance that purifies the tap water that is poured into it.

If there is one thing that bothers people more than unclean water, it is unclear air so it is no wonder there are so many air cleaners and humidifiers available to the public who either have breathing problems or desperately want to avoid them. Some are the electrostatic models that remove up to 99 percent of the air impurities and eliminate smoke and odor at the same time. Others add moisture when needed and eliminate it when not. There is even one that uses an oxygen derivative in the form of an ozone bulb; when the light is turned on, it automatically clears the air. In effect, the light shineth and one can breathe again. How wonderful!

Present

1. Ronson Foodmatic Console, 1973

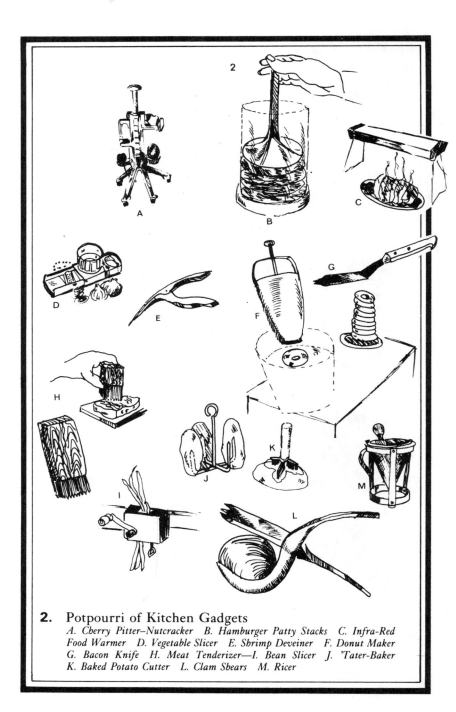

2. Potpourri of Kitchen Gadgets
A. Cherry Pitter–Nutcracker B. Hamburger Patty Stacks C. Infra-Red Food Warmer D. Vegetable Slicer E. Shrimp Deveiner F. Donut Maker G. Bacon Knife H. Meat Tenderizer—I. Bean Slicer J. 'Tater-Baker K. Baked Potato Cutter L. Clam Shears M. Ricer

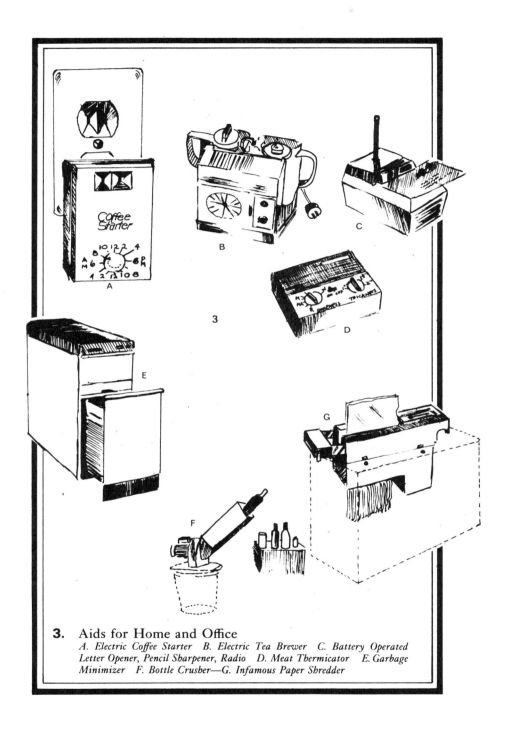

3. Aids for Home and Office
A. Electric Coffee Starter   B. Electric Tea Brewer   C. Battery Operated Letter Opener, Pencil Sharpener, Radio   D. Meat Thermicator   E. Garbage Minimizer   F. Bottle Crusher—G. Infamous Paper Shredder

4. Cappuccino Maker
5. Electric Espresso/Cappuccino Maker

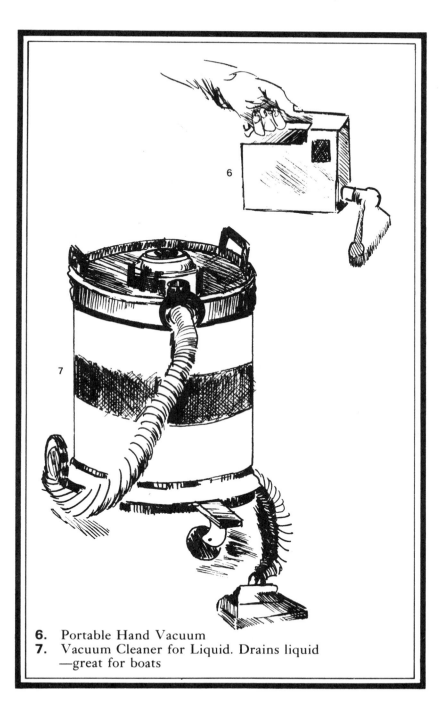

6. Portable Hand Vacuum
7. Vacuum Cleaner for Liquid. Drains liquid —great for boats

8. Do-it-yourself Pasta Maker

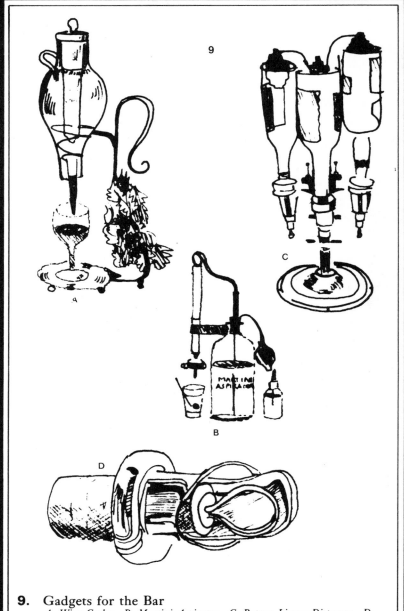

9. Gadgets for the Bar
A. Wine Cooler   B. Martini Aspirator   C. Rotary Liquor Dispenser   D. Dripless Wine Pourer

10. Gadgets for the Car
A. Rear Window Defroster   B. Rear Window Warning Signal   C. Wheel and Brake Lock

11. Gadgets for the Sportsman
A. Folding Saw   B. Swiss Army Knife   C. Folding Bush Knife   D. Fishing Knife

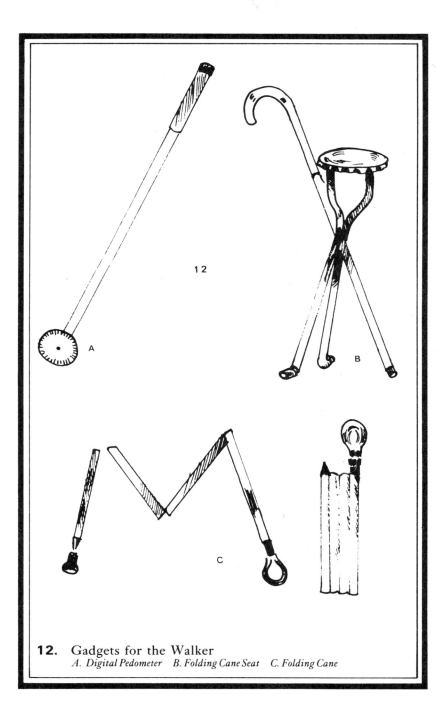

12. Gadgets for the Walker
  A. Digital Pedometer   B. Folding Cane Seat   C. Folding Cane

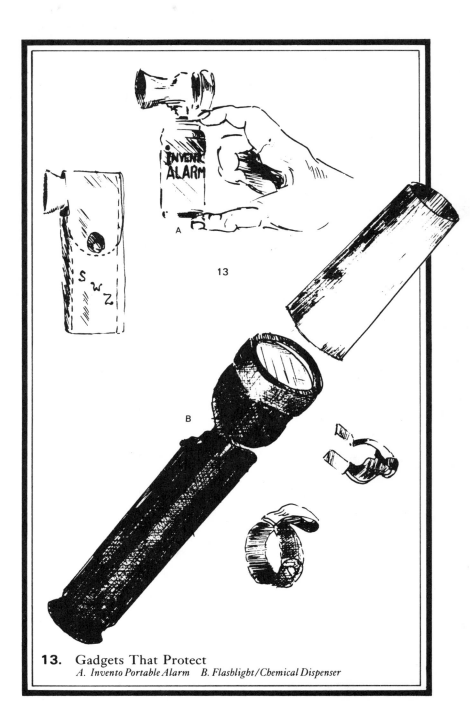

13. Gadgets That Protect
A. Invento Portable Alarm   B. Flashlight/Chemical Dispenser

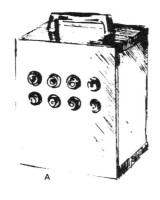
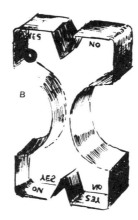

14. Fun and Games
The Nothing Box   B. The Executive Decision Maker   C. The Seduction Box

15. The Better to See You With
 A. Radio Sunglasses   B. No-fog Goggles   C. Glasses With Wipers For The Cyclist   D. TV Glasses. Look up and see straight ahead   E. Fish-Locating Glasses. For looking under water

16. Gadgets for the Hair
A. General Electric Portable Hair Dryer. Cap style, early 1950s B. Hair Dryer Displayed at Patent Expo, 1973

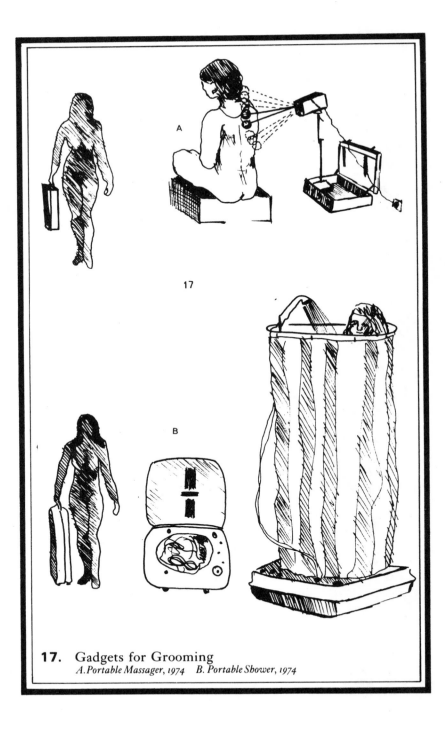

17. Gadgets for Grooming
 A. Portable Massager, 1974   B. Portable Shower, 1974

2001

Gadgets of Fantasy and the Future

What will the gadgets of the future be like? Let's see what the people who create and sell them, the public at large, and the science fiction writers, those fearless forecasters of tomorrow, have to say.

Says Dominic Tampone, president of Hammacher Schlemmer:

> Until today, the rule of thumb has been to make the hand-operated items electrical, and once they're electric, to remove the umbilical cord. But now it's time to go one step further—past the stage of batteries.
>
> I predict that tomorrow's gadgets will operate by the process of induction. A small generator will be plugged into an outlet, but the appliances them-

selves will no longer need to be. They can be set anywhere in the room. When the "On" button is pressed, the appliance will start operating immediately as a result of the high-frequency power transmitted to it by the generator. There will be no end to the number of appliances one generator can service, and the consequences of this are extremely far reaching. It means that finally there will be an end to the ever increasing electricity bill, an end to short circuits caused by overloads of electric current, an end to depending on batteries which invariably go dead at just the wrong time, and an end to unsightly and potentially dangerous cords around the home.

In addition, more and more gadgets will operate via sound waves. For example, with washing machines, sound waves, rather than water, will loosen up the fibers and eliminate the dirt.

One gadget that is sure to live up to its potential in the years to come is the Videovoice. This device operates on the same principle as a radio receiver, except that in this case it permits you to hear the television wherever you are, no matter how far away physically you are from the set. It will be possible, for instance, to turn on the TV evening news at your home with a remote control while you're still driving toward it, and with the Videovoice, to hear the broadcast completely, commercials and all. Closer to home, if you have only one television set, it means you can turn on "The Johnny Carson Show" in the living room and listen to it in the bedroom via an earplug, without disturbing your mate.

When asked what "gadget of tomorrow" he thought most important to develop, Mr. Tampone replied unhesitatingly, "One that will measure the impurities in our air and, when they reach a certain level, get rid of them." Does he see this as a possibility? "If I know anything about the world of gadgets, I know one thing. Anything can happen. And it usually does."

If the gadgets introduced at PatExpo '74, the annual inventor's show in New York, are any barometer of the future, they will share two qualities. They will be disposable and they will be portable.

Many of the newly invented throwaways are aimed at the health field. For dentists, there is a disposable mirror designed to be discarded after one use, which saves having to sterilize it. For squeamish hospital aides, someone has designed a disposable bedpan. And from Mexico comes a syringe that self-destructs after one use, preventing reuse by addicts. There are also disposable neckties, disposable toilet seats, and disposable dustpans.

A problem that this creates, unfortunately, is how to dispose of the disposables. One inventor took a step in the right direction when he came up with a safety pin made from a vegetable-based material. If a child swallows it accidentally, the pin will be safely digested and, for good measure, it is vitamin enriched. But whether or not a dentist can get his patient to swallow his mirror is another question!

The Museum of Contemporary Crafts in Manhattan recently staged an exhibit titled "Portable World" that showed prototypes of gadgets that may well become part of our everyday lives in the years to come. This included such novelties as a "wild oats sowing kit," which included a silver-and-brass pendant containing a Dialpack of contraceptive pills for the woman who likes jewelry and also doesn't know where she may be sleeping that particular evening; a necklace containing a small oxygen mask for those who fear fainting spells; another, trimmed with peacock feathers, that monitors the wearer's body temperature; a stainless-steel belt with a built-in device to monitor pollution in the air for those with respiratory ailments; and an ornate gold-and-silver bracelet containing an electric gadget that measures the pulse rate.

Another portable gadget looking for a manufacturer is Joseph Nebel's "personal breathing mask," a more comfortable and functional version of the old face mask, which can warm up, cool off, and filter the air a person inhales. It can also be adapted for people with a lung or heart ailment by inserting special medication, and for persons working with spray paint or sandblasters, with the use of a filter.

The Paris Inventors Fair in 1974 offered an interesting look at what Europeans are developing and trying to market. Some of the more inventive included a self-blocking, self-tying, shoelace-tying machine; an electric page turner for books; a propeller-operated seltzer-bottle washer (which won seventy-five dollars for its creator); a cigarette holder that keeps ashes from falling off the cigarette; and an apparatus that automatically takes your shoes off.

Major companies, of course, are always researching gadgets of the future. Westinghouse, for example, is currently working on a faucet that turns on and off when you place your hands near it (especially useful when they are covered with paint). Also under development is a bathroom sink with buttons that, when pressed, will regulate the amount and temperature of the water.

On the drawing boards at both Westinghouse and Whirlpool are cylindrical plastic-molded refrigerators with shelves that rotate 360 degrees and eliminate the difficulty of reaching for foods stored in the back of the unit. Whirlpool also envisions ovens and other appliances that will ascend to eye level when a button is pressed, of great value to those who suffer from back problems.

Frigidaire plans to introduce a Touch and Cook electric ranger that features touch controls and relies heavily on computer technology. By touching different points on the Cookmaster panel, the oven and top burners can be programmed to operate and turn off automatically, tell the time, clean itself, and set the temperature for the same food at different times. As an extra attraction, a preheat cycle can be automatically programmed so that food can be kept warm for late arrivals.

A number of new watches with new gadgets are now in production, among them, Patek Phillipe's calendar model that automatically allows February a twenty-ninth day every four years—for a mere $4,500. Not to be outdone, Bulova is developing a radio watch that will permit its wearer to tell the time on any planet of his choice, providing he doesn't mind wearing an antenna on his head through which he'll receive the information.

Happily, the trend toward safety and comfort in automotive gadgets is continuing. It is expected that standard equipment on all cars will soon include rear lights that will automatically intensify according to the daylight, fog, etc.; seats fixed and welded to side pillars and a center tunnel to form a transverse bulkhead; steering columns and pedals adjusted with the push of a button to accommodate drivers of any size, single-blade wipers that move in the direction of the airstream, reducing the tendency for the wiper to lift off the glass in a storm; and leak-proof fuel tanks.

Automobile gadgets were also the highlight of the Inventor's Fair in Paris. André Goury came up with one that fires a rifle shot when the door of a parked car is forced open. And for those who prefer a solution a little less violent, the rifle shot is replaced with a firing tube that will "send out some object that makes a lot of noise when it falls" or one that could tip over a container of some noxious liquid or powder that would adhere to the would-be perpetrator. Joseph Corporandy demonstrated a gadget that automatically cuts off the battery current in case of a severe jolt, reducing the danger of fire in an accident.

· And lastly, depending on your point of view, a boon for those who may get stuck in a traffic jam, a writing table that fits over the steering wheel, which thereby deprives the driver of an excuse for not getting his paper work done on time.

Practicality notwithstanding, what kind of gadgets would the general public like to find? *Science and Mechanics* magazine, among others, has a monthly column where readers' suggestions are solicited. Some of the responses have illustrated a desire for such items as:

Shopping carts with steering wheels

Bumper stickers that self-destruct when an election is over

Edible tape for holding big sandwiches together

Bio-degradable campaign signs

Fish hooks with a camera on them to take a picture of "the one that got away"

Padded bottoms on ketchup bottles so people won't hurt their hands trying to get the ketchup out

Sugar-coated spoons for taking medicine

Hammers with glass heads so that people trying to get out their frustrations with them can't break anything but the hammer

Sun-magnifying parasols covered with a high-powered magnifying glass so that on a sunny day, you can get a tan in half the time

Ice-cream-cone holders, with covers to keep them cold, so you can eat them at home rather than slurp them on the street

Contour cue-ball combers with teeth at each end to comb sideburns for baldies

Tiny Tim keys for short people with high locks on doors or tall people with short locks

A slot in the car's engine for keeping pizza warm until it gets home

Forks with cranks to turn the prongs for eating spaghetti

Ice skates with turn signals

Drinking straws with a pump for milkshakes.

Other suggestions from those who, from time to time, dwell in fantasyland, include:

A truth filter for political campaign speeches

Edible light bulbs for people who want a "light snack"

Safety nets for people who jump to conclusions

A toaster bed so when it's time to get up, you pop out of bed

Retractable pens for people who write letters of apology

A dieter's fork made in such a way it absolutely prevents the lifting of food

A combination record player and air conditioner for people who play it cool

An extra lid for people who blow their tops

A knife-and-dryer combination for those who like things cut and dried.

Science fiction writers have always had a special expertise in predicting what is going to happen before it happens. After all, they were the ones who described television as far back as 1919, 3-D films, computers (which they referred to then as mechanical brains), radar, robots, rockets, and man's space journey to the moon. Science fiction has truly become science fact.

If one were to believe Aldous Huxley, gadgets in tomorrow's society will be exceptionally complicated mechanical versions of today's simplest form of enjoyment. In *Brave New World*, he introduces the Centrifugal Bumble Puppy, a children's game in which a ball is thrown up to a platform on the top of a tower. The ball then rolls down into a rapidly revolving disc located in the interior, is hurled through one of the many holes pierced in the cylindrical casing, and is finally caught by the "lucky" kid—in effect, a mechanized game of catch. In fact, in the Brave New World, the Controllers won't even give approval to any new game unless it requires at least as much apparatus, if not more so, as the most complicated of existing games.

George Orwell prophesied the paper shredder as far back as 1949 when he wrote *1984*. His shredders consisted of orifices in the walls that were nicknamed "memory holes." Whenever one finished reading a piece of paper, he would drop it in, whereupon it would be whirled away on a current of warm air to be burned in an enormous furnace hidden somewhere in the recesses of the building.

More frightening, he predicted a life that will be directed by a voice behind the "telescreen," a kind of television that can be lowered but never turned off, so sensitive it can pick up the beat of a viewer's heart and monitor his every move. The voice talks to each viewer personally, gives exercise classes each morning, chastises the citizen who does not give 100 percent of himself, and dispenses ideology throughout the day and night. It is at moments like this one hopes science fiction writers might be wrong.

On the lighter side, of course, there's always Woody Allen's Orgasmotron, a box that one enters to receive the thrill of a lifetime, which he so vividly demonstrated in his hilarious film of life in 2173 A.D., *Sleeper*. And there's Fred Pohl's Joymaker, which he describes in *The Age of Pussyfoot*. The Joymaker is a gadget that "awakes one in the morning and induces sleep at night. It feeds and clothes and delivers messages. It administers tranquilizers or stimulants whenever it thinks necessary and it can 'spray' one with a variety of sensations, each more delightful than the last."

Will these come to pass, for better or for worse? Only time will tell.

Obviously, no one knows for sure what the gadgets of tomorrow will be like. But one thing is certain. They, like their counterparts today, will reflect the life-styles, needs, and fantasies of the generations that create and consume them. And, as has been the case throughout the ages, they will be practical and impractical, disposable and indisposable, useful and useless, expensive and inexpensive, simple and complicated, dispensable and indispensable, lightweight and heavy, relevant and irrelevant, amusing, fun, imaginative, innovative, unique, ingenious, and often incredible.

GADGETS will always be GADGETS.

Future

1. Sonic Sewing Machine

2. Gadgets That Make Iron Man Iron Man
Marvel Comics Group

3. Automatic Shoe Remover
Gadget can remove shoes and tie laces

4. Gadget That Notifies Burglee and Burglar

5. Whimsey Mobile
Post-Rube Goldberg

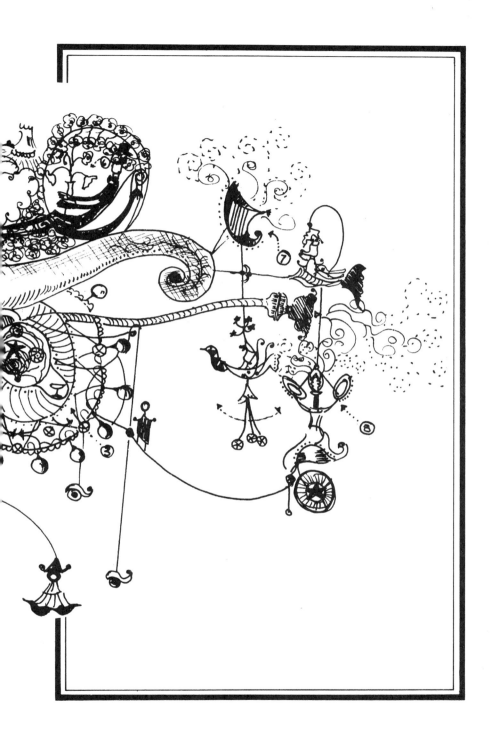

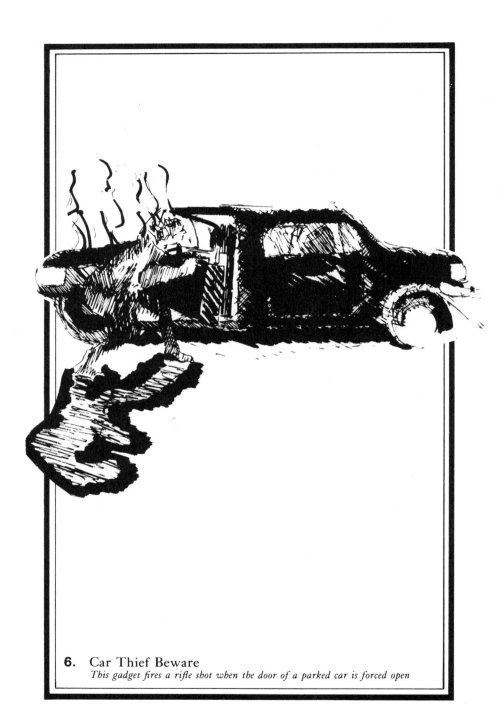

6. Car Thief Beware
This gadget fires a rifle shot when the door of a parked car is forced open

WHERE TO FIND THEM
Resource Section

HOUSEHOLD/LOCKS/SAFETY

Antimugging shock rods. Available Sunset House, 187 Sunset Building, Beverly Hills, California 90201. $7.99.

Automatic appliance timer. Intermatic Time-all Timer. Available Macy's, Herald Square, New York City. $9.99.

Automatic paint roller. Available Bevi Industries, Westmoreland Avenue, White Plains, New York. 10606. $4.98.

Furnace alarm. Available Edmund Scientific Company, 300 Edscorp Building, Barrington, New Jersey. $12.25.

Inside-outside window cleaner. Magna-Clean. Available Hammacher Schlemmer, 141 East 57th Street, New York City. $9.95.

Instant safety ladder. Available Hammacher Schlemmer, 141 East 57th Street, New York City. Fifteen feet, $19.95. Twenty-five feet, $30.00.

Intruder door lock and alarm. Available American Consumer, 195 Shippan Avenue, Stamford, Connecticut. 06904. $4.98.

Multipurpose alarm. Available Edmund Scientific Company, 300 Edscorp Building, Barrington, New Jersey. $35.95.

Remote electronic thermometer. Available James Electronics, 4050 North Rockwell Street, Chicago, Illinois 60618. $39.95.

Remote temperature check. Dial-Temp. Available Slawek Enterprises, P.O. Box 5811, Philadelphia, Pennsylvania 19128. $39.95.

Self-coiling cord. Available Thomas Sales and Marketing, Bloomington, Minnesota. $1.98.

Smoke-Gard. Available Island Research Company, 115 Robby Lane, Manhasset, New York 11040. $45.95.

Ultrasonic intruder alarm. Available Lafayette Electronics, 111 Jericho Turnpike, Syosset, New York 11791. $59.95.

Horn alarm. Invento. Available Hammacher Schlemmer, 141 East 57th Street, New York City. $10.00.

Swiss Army knife. Available Hoffritz, 20 Cooper Square, New York City. $28.95.

KITCHEN GADGETS

Apple parer/corer/slicer. Hand-operated. Available Bazar Français, 666 Sixth Avenue, New York City. $7.50.

Bacon grill. Available Hammacher Schlemmer, 141 East 57th Street, New York City. $8.50.

Baked-potato puffer. Available Colonial Garden Kitchen, 270 West Merrick Road, Valley Stream, New York 11580. $1.19.

Boeuf à la mode needles. Available Bazar Français, 666 Sixth Avenue, New York City. Sixteen-inch, $7.00. Eighteen-inch, $8.00.

Chestnut pan. Available Bazar Français, 666 Sixth Avenue, New York City. $3.00.

Combination screwdriver/pliers/hammer/triple wrench. Everything Tool. Available House of Value, 2052 Albatross, San Diego, California. $2.98.

Corn-kernel remover. Available Bazar Français, 666 Sixth Avenue, New York City. $2.40.

Couscous cooker. Available Bazar Français, 666 Sixth Avenue, New York City. $18.00.

Croissant cutter. Available Bazar Français, 666 Sixth Avenue, New York City. $22.00.

Double-boiler support. Available Bazar Français, 666 Sixth Avenue, New York City. $2.25.

Duck press. Available Bazar Français, 666 Sixth Avenue, New York City. $250.00.

Egg pick. Available Lillian Vernon, 510 South Fulton Street, Mount Vernon, New York 10550. $1.00.

Electric brisker. Available Hammacher Schlemmer, 141 East 57th Street, New York City. $25.00.

Electric preset coffee starter. Available Hammacher Schlemmer, 141 East 57th Street, New York City. $7.95.

Electric egg cooker/poacher. Available Hammacher Schlemmer, 141 East 57th Street, New York City. $24.95.

Electric egg inspector. Available Hammacher Schlemmer, 141 East 57th Street, New York City. $8.95.

Electric food smoker. Available Hammacher Schlemmer, 141 East 57th Street, New York City. $18.00.

Electric food warmer. Invento Prima Infra-Red Food Warmer. Available Hammacher Schlemmer, 141 East 57th Street, New York City. $18.00.

Electric Foodmatic. Ronson. Available major department stores. $190.00.

Electric fruit juicer. Fresh Vitamin C Fruit Juicer. Available Hammacher Schlemmer, 141 East 57th Street, New York City. $27.50.

Electric hot-dog cooker. Hot Dogger. Available Hammacher Schlemmer, 141 East 57th Street, New York City. $9.95.

Electric pepper mill. Invento. Available Hammacher Schlemmer, 141 East 57th Street, New York City. $12.95.

Electric potato peeler. Not available at this time.

Electric salad maker. Moulinex. Available Bazar Français, 666 Sixth Avenue, New York City. $35.00.

Electric steak timer. Steakwatcher. Available Hammacher Schlemmer, 141 East 57th Street, New York City. $24.95.

Electric Teasmade. Available Hammacher Schlemmer, 141 East 57th Street, New York City. $139.50.

Electric vegetable juicer. Available Hammacher Schlemmer, 141 East 57th Street, New York City. $45.00.

Fish grill. Available Hoffritz, 20 Cooper Square, New York City. $7.50.

Frypan cover. Available Sunset House, 187 Sunset Building, Beverly Hills, California 90201. $1.49.

Heat diffuser. Available Bazar Français, 666 Sixth Avenue, New York City. $2.50.

Juice-can opener. Liqui-Pour. Available Walter Drake, Drake Building, Colorado Springs, Colorado 80915. 98¢.

Ketchup Saver. Available Colonial Garden Kitchen. 270 West Merrick Road, Valley Stream, New York 11580. $1.49.

King Roast Pin. Available Hammacher Schlemmer, 141 East 57th Street, New York City. $14.95.

Knife sharpener. Dial-X. Available Bevi Industries, Westmoreland Avenue, White Plains, New York 10606. $12.50.

Meat injector. Flavor Jector. Not available at this time.

Oil and vinegar vials. Available Hammacher Schlemmer, 141 East 57th Street, New York City. $17.95.

Omelet lifter. Available Bazar Français, 666 Sixth Avenue, New York City. $1.25.

Pop-It Burger. Available Colonial Garden Kitchen, 270 West Merrick Road, Valley Stream, New York. $1.95.

Potato baker. Tater-Baker. Available Walter Drake, Drake Building, Colorado Springs, Colorado 80915. 98¢.

Pastry blender. Available Bazar Français, 666 Sixth Avenue, New York City. $1.75.

Pepper mill, fish shape. Available Hammacher Schlemmer, 141 East 57th Street, New York City. $25.00.

Pineapple corer. Available Hammacher Schlemmer, 141 East 57th Street, New York City. $2.95.

Poultry clips. Available Colonial Garden Kitchen, 270 West Merrick Road, Valley Stream, New York. $2.49.

Rice cooker. Available Bazar Français, 666 Sixth Avenue, New York City. $2.75.

Roast rack, adjustable. Available Bazar Français, 666 Sixth Avenue, New York City. $3.00.

Rolling pin, three discs. Available Colonial Garden Kitchen, 270 West Merrick Road, Valley Stream, New York. $4.95.

Salad dryer. Available Hammacher Schlemmer, 141 East 57th Street, New York City. $12.50.

Salad tosser set. Available Hammacher Schlemmer, 141 East 57th Street, New York City. $6.50.

Salt-and-pepper set, automatically clean. Available Colonial Garden Kitchen, 270 West Merrick Road, Valley Stream, New York. $3.98.

Self-stirring saucepan. Available Hammacher Schlemmer, 141 East 57th Street, New York City. $34.95.

Syrup dispenser. Invento. Available Hammacher Schlemmer, 141 East 57th Street, New York City. $9.50.

Third Arm. Reach-It. Available Colonial Garden Kitchen, 270 West Merrick Road, Valley Stream, New York. $3.75.

Toast warmer. Available Hammacher Schlemmer, 141 East 57th Street, New York City. $15.00.

Tri-fry pan. Available Hammacher Schlemmer, 141 East 57th Street, New York City. $9.00.

Weight Ladle. Available Bazar Français, 666 Sixth Avenue, New York City. $11.00.

BAR GADGETS

Automatic martini mixer. Autobar. Available Hammacher Schlemmer, 141 East 57th Street, New York City. $395.00.

Brandy warmer. Available Hoffritz, 20 Cooper Square, New York City. $9.00.

Champagne recorker. Available Hammacher Schlemmer, 141 East 57th Street, New York City. $10.00.

Champagne tap. Available Hammacher Schlemmer, 141 East 57th Street, New York City. $7.50.

Classic syphon. Available Hammacher Schlemmer, 141 East 57th Street, New York City. $25.00.

Corklifter. Available Hoffritz, 20 Cooper Square, New York City. $6.95.

Dripless wine pourer. Available Hammacher Schlemmer, 141 East 57th Street, New York City. $7.50.

Spill-Gard ice tray. Available Hammacher Schlemmer, 141 East 57th Street, New York City. $3.00.

GlassMaker. Available Hammacher Schlemmer, 141 East 57th Street, New York City. $29.95.

Rotary liquor dispenser. Available Hammacher Schlemmer, 141 East 57th Street, New York City. $195.00.

Wine frigidarium. Available Hammacher Schlemmer, 141 East 57th Street, New York City. $15.00.

Austrian drink cooler. Available Hammacher Schlemmer, 141 East 57th Street, New York City. $25.95.

Cold can popper-upper. Available Hammacher Schlemmer, 141 East 57th Street, New York City. $16.95.

Calibrated martini dropper. Available Hammacher Schlemmer, 141 East 57th Street, New York City. $3.95.

HEALTH AND MEDICAL

Bio-feedback trainer. Available Edmund Scientific Company, 300 Edscorp Building, Barrington, New Jersey 08007. $120.00.

Cigarette-smoking reducer. Dial-Down. Available Bevi Industries, Westmoreland Avenue, White Plains, New York 10606. $2.98.

Dent-U-Sonic denture cleaner. Available Hammacher Schlemmer, 141 East 57th Street, New York City. $10.00.

Exerciser Exer-Gym. Isotonic/Isometric. Available Haverhill's, 584 Washington Street, San Francisco, California 94111. $26.95.

Jogging board. Executive Jogger. Available Hammacher Schlemmer, 141 East 57th Street, New York City. $9.95.

Hydraulic dental appliance. Family Dental Kit. Available American Consumer, 741 Main Street, Stamford, Connecticut 06904. $7.98.

Figuretone exercise bike. Not available at the moment.

Lighted medical spoon. Available Gus File, Inc., Box 3006, Albuquerque, New Mexico 87110. $1.95.

Sleep Sound. Available Hammacher Schlemmer, 141 East 57th Street, New York City $22.95. *Surf and Rain Sleep Sound,* $75.00.

Snore stopper. Naso-Vent. Available Things Et-Cetera, 210 Fifth Avenue, New York City. $2.95.

Solid Oxygen System. Available Dow Chemical Company, Midland, Michigan. $295.00.

Sphygmostat electronic blood-pressure monitor. Available Hammacher Schlemmer, 141 East 57th Street, New York City. $120.00.

Toothpaste dispenser. Spenzal Tube Dispenser. Available Dapuzzo Sales, 191–16 35th Avenue, Flushing, New York 11358. $2.25.

Electronic thermometer. Available Hammacher Schlemmer, 141 East 57th Street, New York City. $45.00.

Always Empty ashtray. Hammacher Schlemmer, 141 East 57th Street, New York City. $20.00

Where to Find Them

GROOMING

Beard-and-mustache groomer kit. Available Hammacher Schlemmer, 141 East 57th Street, New York City. $29.95.

Electric heated towel stand. Available Hammacher Schlemmer, 141 East 57th Street, New York City. Standard model, $79.50. Deluxe model, $89.50.

Hair detangler. Gillette. Available major department stores. $19.95.

Hindsight coiffure mirror. Available Hammacher Schlemmer, 141 East 57th Street, New York City. $10.00.

Neck mirror. Available Sunset House, 187 Sunset Building, Beverly Hills, California 90201. $1.00.

Posture developer. Belly Beeper. Available Bevi Industries, Westmoreland Avenue, White Plains, New York. $9.98.

Razor-blade sharpener. Span-O-Matic. Available Hammacher Schlemmer, 141 East 57th Street, New York City. $5.95.

Shower-head jet spray. Available Hammacher Schlemmer, 141 East 57th Street, New York City. $17.95.

Whirlpool bath attachment. Aquajet. Available Jay Norris Corporation, 25 West Merrick Road, Freeport, New York 11520. $4.99.

Electric presser. Invento. Available Hammacher Schlemmer, 141 East 57th Street, New York City. $150.00.

Electric tie rack. Available Hammacher Schlemmer, 141 East 57th Street, New York City. $9.00.

Pair-O-Socks. Available Walter Drake, Drake Building, Colorado Springs, Colorado 80915. 98¢

Boot-'n-shoe brush. Available Hammacher Schlemmer, 141 East 57th Street, New York City. $18.95.

Turbo-jet for bath. Available Hammacher Schlemmer, 141 East 57th Street, New York City. $180.00.

Stably Razor. Available Hammacher Schlemmer, 141 East 57th Street, New York City. $29.95.

OFFICE GADGETS

Automatic electric speaker telephone. Available Grand Com, 324 Fifth Avenue, New York City. $250.00.

Automatic telephone recorder switch. Tele-A-Corder. Available Lafayette

Radio Electronics, 111 Jericho Turnpike, Syosset, New York. $25.95.

Calculator pencil. Available Johnson Smith Company, 35075 Automation Drive, Mount Clemens, Michigan 48043. $5.95.

Checkbook adding machine. Dial-A-Balance. Available Jay Norris Corporation, 25 West Merrick Road, Freeport, New York 11520. $2.99.

Conference-call telephone device. Tele-Caller. Available Lafayette Radio Electronics, 111 Jericho Turnpike, Syosset, New York. $24.95.

Cordless extension telephone. Dial-O-Matic Pocket Fone. Available Dial-O-Matic, 339 Broadway, New York City. $48.00.

Counterfeit-bill detector. Available Goldsmith Brothers, 77 Nassau Street, New York City. $26.95.

Electric heat pad. Radiant Heat Pad. Available Goldsmith Brothers, 77 Nassau Street, New York City. $17.50.

Electric letter opener. Available Hammacher Schlemmer, 141 East 57th Street, New York City. $17.95.

Electric paper shredder. Available Michaels Business Machines, 145 West 45th Street, New York City. $2,400.00.

Electric pencil sharpener. Apscomatic. Available Goldsmith Brothers, 77 Nassau Street, New York City. $22.95.

Electric tape dispenser. Available Bullock and Jones, 340 Post Street, San Francisco, California 94108. $7.95.

Hush-a-phone. Available Bevi Industries, Westmoreland Avenue, White Plains, New York 10606. $12.95.

Left-handed tape measure. Available The Left Hand, 145 East 27th Street, New York City. $12.95.

Multilanguage typewriter. Interlingual Typewriter. Available Interlingual Cultural Machinery, 94 Elizabeth Street, New York City. $400.00.

Pocket calculator. Sinclair. Available Liberty Music Store, 450 Madison Avenue, New York City. $195.00.

Remote telephone recorder attachment. Remote Mate. Available Hammacher Schlemmer, 141 East 57th Street, New York City. $129.95.

Phone Mate. Available Hammacher Schlemmer, 141 East 57th Street, New York City. $119.00.

Self-correcting typewriter ribbon. Mail-order Mart, 2701 Sterlington Road, Monroe, Louisiana 71201. $3.50.

Stapleless Paper Stapler. Available Hammacher Schlemmer, 141 East 57th Street, New York City. $15.50.

Where to Find Them

Tape dispenser. Wrist Tape-Toter. Available Alpine Valley Enterprises, 138 Maple Street, Wilmington, Delaware 19808. $1.29.

Tape measure. All-in-One Tape. Available Bullock and Jones, 340 Post Street, San Francisco, California 94108. $10.00.

Telephone amplifier. Available Lafayette Radio Electronics. 111 Jericho Turnpike, Syosset, New York. $12.95.

Telephone-call transfer service. Divert-A-Matic. Available Telephone Improvement Service, 160 West 46th Street, New York City. 83¢ per day.

Telephone Name Caller. Available Goldsmith Brothers, 77 Nassau Street, New York City. $49.95.

Telephone-cap hold device. Hold-On. Available Hoffritz, 20 Cooper Square, New York City. $3.95.

Telephone time limiter. Teen Tel Limiter. Available Lafayette Radio Electronics, 111 Jericho Turnpike, Syosset, New York. $29.95.

TRAVEL GADGETS

Automatic fish feeder. Available Hammacher Schlemmer, 141 East 57th Street, New York City. $15.95.

Automatic watering flower-pot holder. Available Hoffritz, 20 Cooper Square, New York City. $15.95.

Converter-adapter. Available Hoffritz, 20 Cooper Square, New York City. $10.95.

Travel iron. Available Hammacher Schlemmer, 141 East 57th Street, New York City. $12.95.

World's smallest alarm clock. Available Mark Cross, 55th Street and Fifth Avenue, New York City. $65.00.

Folding cane. Available Hammacher Schlemmer, 141 East 57th Street, New York City. $8.95.

Folding wig stand. Available Walter Drake, Drake Building, Colorado Springs, Colorado 80915. $2.95.

Immersion heater. Available Hammacher Schlemmer, 141 East 57th Street, New York City. $8.95.

Luggage wheels. Baggage Master. Available Hammacher Schlemmer, 141 East 57th Street, New York City. $10.00.

Miniature wrinkle remover. Available Hoffritz, 20 Cooper Square, New York City. $19.95.

Electric barber. Available Hammacher Schlemmer, 141 East 57th Street, New York City. $32.50.
Electric hair styler. Available Hammacher Schlemmer., 141 East 57th Street, New York City. $22.00.
Pocket pepper mill. Available Hoffritz, 20 Cooper Square, New York City. $10.00.
Steam/press valet. Available Hammacher Schlemmer, 141 East 57th Street, New York City. $12.95.
World-Wide immersion heater. Available Hoffritz, 20 Cooper Square, New York City. $12.95.
Hair dryer. Braun International. Available Hammacher Schlemmer, 141 East 57th Street, New York City. $18.95.

AUTOMOBILE GADGETS

Air-cooled seat. Invento. Available Hammacher Schlemmer, 141 East 57th Street, New York City. $15.00.
Auto antenna light. Available Worldwide, 426 Washington, Oak Park, Illinois 60302. $1.49.
Auto beverage caddy. Available J. C. Whitney, 1917 Archer Avenue, Chicago, Illinois 60616. $1.59.
Auto change dispenser. Available Sunset House, 187 Sunset Building, Beverly Hills, California 90201. $1.00.
Auto defroster gun. Invento. Available Hammacher Schlemmer, 141 East 57th Street, New York City. $10.00.
Auto hot seat. Available Johnson Smith Company, 35075 Automation Drive, Mount Clemens, Michigan 48043. $2.95.
Auto radar sentry. Available Radatron Corporation, P.O. Box 177, North Towanda, New York. $39.95.
Auto vacuum. Available Hammacher Schlemmer, 141 East 57th Street, New York City. $17.95.
Auto warning system. Available Hammacher Schlemmer, 141 East 57th Street, New York City. $9.95.
Electric car. Available Eagle Pitcher Industries, Detroit, Michigan. $4,000 for batteries.
Remote-control garage-door opener. Alliance Genie. Available Macy's, Herald Square, New York City. $119.95.
Turnpike toll gun. Available Hammacher Schlemmer, 141 East 57th Street, New York City. $9.95.

Where to Find Them

FUN, GAMES, AND SPORTS

All-in-one golf club. Metropole. Available Abercrombie & Fitch, Madison and 45th Street, New York City. $42.00.

Automatic card shuffler. Invento. Available Hammacher Schlemmer, 141 East 57th Street, New York City. $16.95.

Automatic golf score keeper. Available Hoffritz, 20 Cooper Square, New York City. $8.50.

Automatic poker machine. Available Hammacher Schlemmer, 141 East 57th Street, New York City. $12.95.

Automatic range finder. Available Abercrombie & Fitch, Madison and 45th Street, New York City. $34.95.

Bleeping golf-ball finder. Available Hammacher Schlemmer, 141 East 57th Street, New York City. $30.00.

Combination golf ball/shoe cleat/iron groove cleaner. Available Elgin Engraving Company, 970 Edwards Avenue, Dundee, Illinois 60118. $2.95.

Electronic mosquito chaser. Available Hanover House, Hanover, Pennsylvania 19331. $9.98.

Electronic metal detector. Viking Snooper-Tronic. Available Hammacher Schlemmer, 141 East 57th Street, New York City. $29.95.

Executive decision maker. Available Hammacher Schlemmer, 141 East 57th Street, New York City. $10.95.

Fish cleaning board. Available Abercrombie & Fitch, Madison and 45th Street, New York City. $8.95.

Fish skinner. Available Abercrombie & Fitch, Madison and 45th Street, New York City. $5.95.

Fishing net. Available Abercrombie & Fitch, Madison and 45th Street, New York City. $8.95.

Golf-ball heater. Fireball. Available Box 2355, Palos Verdes, California 90274. $14.95.

Golf practice range. Available Abercrombie & Fitch, Madison and 45th Street, New York City. $29.95.

Golf swing trainer/exercise club. Available Abercrombie & Fitch, Madison and 45th Street, New York City. $16.95.

Hot-weather hat. Not available at this time.

Locator for fish. Lowrance Fish Lo-K-Tor. Available Abercrombie & Fitch, Madison and 45th Street, New York City. $159.95.

Nothing Box. Available Hammacher Schlemmer, 141 East 57th Street, New York City. $29.95.

Seat cane. Available Sunset House, 187 Sunset Building, Beverly Hills, California. $9.99.

Ski speedometer. Skidometer. Available The Gallery, Department 6792, Amsterdam, New York 12010. $10.00.

Sonus Sound Switch. Available Hammacher Schlemmer, 141 East 57th Street, New York City. $29.95.

Target launcher. Available Abercrombie & Fitch, Madison and 45th Street, New York City. $15.00.

Target practice light. Spot-Shot Ltd. Available Abercrombie & Fitch, Madison and 45th Street, New York City. $49.50.

Tennis rebound net. Available Abercrombie & Fitch, Madison and 45th Street, New York City. $55.00.

Parking-meter lamp. Available Hammacher Schlemmer, 141 East 57th Street, New York City. $50.00.

Video Voice. Available Hammacher Schlemmer, 141 East 57th Street, New York City. $29.95.

DO-IT-YOURSELF/ECOLOGY

Drain declogger. Mini-Jet. Available Bevi Industries, Westmoreland Avenue, White Plains, New York. $2.98.

Electric paint remover. Available Bevi Industries, Westmoreland Avenue, White Plains, New York. $14.95.

Electric refrigerator defroster. Defrost-Fast. Available Edmund Scientific Company, 300 Edscorp Building, Barrington, New Jersey. $4.99.

Noise-pollution meter. Available Edmund Scientific Company, 300 Edscorp Building, Barrington, New Jersey. $76.00.

Spaetzle maker. Available Hammacher Schlemmer, 141 East 57th Street, New York City. $8.50.

Electric Home Cheesery. Make-your-own. Available Hammacher Schlemmer, 141 East 57th Street, New York City. $12.95.

Neapolitan coffee mill. Available Hammacher Schlemmer, 141 East 57th Street, New York City. $109.50.

Make-your-own-fresh-creamer. Available Hammacher Schlemmer, 141 East 57th Street, New York City. $11.95.

Electric ice-cream freezer. Available Hammacher Schlemmer, 141 East 57th Street, New York City. $28.50.

151 • Where to Find Them

Pasta maker. Available Hammacher Schlemmer, 141 East 57th Street, New York City. $38.95.

Natural yogurt maker. Available Hammacher Schlemmer, 141 East 57th Street, New York City. $19.95.

Bread maker. Available Hammacher Schlemmer, 141 East 57th Street, New York City. $26.95.

LIST OF CATALOGS

Abercrombie & Fitch. Madison and 45th Street, New York, New York.

Bazar Français. 666 Sixth Avenue, New York, New York.

Bevi Industries. Westmoreland Avenue, White Plains, New York.

Bullock and Jones. 340 Post Street, San Francisco, California.

Colonial Garden Kitchens. 27010 Merrick Road, Valley Stream, New York 11582.

Catalogue of the Unusual. by Harold Hart. Hart Publishers, 719 Broadway, New York, New York.

Edmund Scientific Company. 300 Edscorp Building, Barrington, New Jersey 18087.

Goldsmith Brothers. 77 Nassau Street, New York, New York.

Hammacher Schlemmer. 141 East 57th Street, New York, New York.

Hoffritz, Inc. 20 Cooper Square, New York, New York.

House of Values. 2052 Albatross, San Diego, California 92101.

Invento. 205 East Post Road, White Plains, New York 10601.

Jay Norris. 25 West Merrick Road, Freeport, New York.

Johnson Smith Company. 35075 Automation Drive, Mount Clemens, Michigan 48043.

Lillian Vernon Company. 510 South Fulton Street, Mount Vernon, New York.

The Left Hand. 145 East 27th Street, New York, New York.

Macy's. Herald Square, New York, New York.

Sunset House. 187 Sunset Building, Beverly Hills, California 90213.

Walter Drake & Sons. Drake Building, Colorado Springs, Colorado 80915.

Worldwide. 426 East Washington, Oak Park, Illinois 60302.

INDEX

Abacus, 8
Abercrombie & Fitch, 3
Abercrombie & Fitch (catalog), 151
Age of Pussyfoot, The (Phol), 130
Air Call, 78
Air cleaners, 106
Air-cooled seat, 86, 148
Alert (Alcohol Level Evaluation Road Tester), 84
Allen, Woody, 130
Alliance Genie, 87, 148
All-in-one golf club, 99, 149
All-in-One Tape, 75, 147
Always Empty ashtray, 144
American Revolution, 15, 21
Antimugging shock rods, 38–39, 139
Appert, Nicholas, 19
Apple parer, 42

Apple parer/corer/slicer, 140
Appliances, turning on (remote control), 39
Apscomatic electric sharpener, 75, 146
Aquafilters, 58
Aquajet, 67,
AquaMassage Shower Head, 66–67
Arbuckle, John and Charles, 21
Ardate Company, 91
Asparagus cooker, 49
Astra-Lite, 80, 146
Austrian drink cooler, 144
Auto antenna light, 148
Auto beverage caddy, 86, 148
Auto change dispenser, 86, 148
Auto defroster gun, 85, 148
Auto-Guardian, 83

Auto hot seat, 87, 148
Auto radar sentry, 83–84, 148
Auto-Va, 86–87, 148
Auto warning system, 83, 148
Autobar, 143
Automatic appliance timers, 39–40, 139
Automatic card shuffler, 96–97, 149
Automatic electric speaker phone, 79, 145
Automatic fish feeder, 90, 147
Automatic golf score keeper, 99, 149
Automatic martini mixer, 53, 143
Automatic paint roller, 41, 139
Automatic Plant Quencher, 5–6, 90, 147
Automatic poker machine, 97, 149
Automatic range finder, 98–99, 149
Automatic telephone recorder switch, 145–146
Automatic watering flower-pot holder, 90, 147
Automobile items, 82–89, 127
    where to find, 148

Babylon (ancient), 8
Bacon grill, 46, 140
Baggage Master, 92, 147
Baked-potato puffer, 47, 140
Banking services after hours, 80–81
Bar gadgets, 52–56
    where to find, 143–144
Barbed-wire stretcher, 12
Barber comb, 69
Barbi (Baseband Radar Bag Initiator), 84
Bath security rail, 67
Bathroom sink, 126
Bathtubs, 10, 15–16
Bazar Français (catalog), 151
Bazar Français (store), 4, 6, 48
Beach, Chester A., 20

Beard-and-mustache groomer, 69, 145
Beaters, 18
Beepers, telephone, 77, 78
Bell Laboratories, 79
Belly Beeper, 61–62, 145
Bevis Industries (catalog), 151
Bio-degradable campaign signs, 128
Bio-feedback trainer, 65, 144
Bissell, Mel, 16
Bleeper Electronic golf ball, 99 149
Blender, 17, 20–21
Blender Cookbook, the, 21
Boeuf à la Mode needles, 45, 140
Boilmaster, 47
Bonnet dryer, 70
Book shelves, 10
Boot husker, 71
Boot shaper, 71
Boot-'n-Shoe brush, 71, 145
Borch, Fred, 60
Borgia, Cesare, 9
Brandy warmer, 56, 143
Braun International Hair Styler, 91
Brave New World (Huxley), 129
Bread makers, 103, 151
Breakmatic, 56
Breathalyser, 84
Brezhnev, Leonid, 58
Bronze Age, 7
Bullock and Jones (catalog), 151
Bulova Watch Company, 127
Bumper stickers that self-destruct, 128
Butter curlers, 5
Butter cutters, 49
Butter spreader-dispenser, 50
Button machine, 14

Cabbage graters, 43
Calculator pencil, 74, 146
Calibrated martini dropper, 53, 144

Can opener, 4, 17–18
Car defroster gun, 85, 148
Card shuffler, 96–97
Cardiomed, 62–63
Carpet sweeper, 16
Cartridge ribbons, 72
Cascade, 93–94
Cassette telephones, 79
Cast iron nutcrackers, 42
Catalogs, list of, 151
Catalogue of the Unusual (Hart), 151
Cattle brand rings, 12
Chafing dish, 43
Chain lock, 37
Champagne Cork Tapper, 55–56, 143
Checkbook adding machine, 76, 146
Champagne recorker, 56, 143
Cheese grater, 43
Cheese slicer, 43
Chemex coffee maker, 22
Cherry pitter, 49
Chestnut pan, 46, 140
Child, Julia, 6
Cigar cutter, 12, 73
Cigarette holder-clamp, 58
Cigarette lighters, 59–60
Cigarette-smoking reducer, 57–58, 144
Classic syphon, 143
Closet tub, 15
Coffee grinder, 13, 42
Coffee maker, 21–22
Coffee mill, 21, 150
Coffee pot, 21–22
Cold can popper-upper, 144
Collision sensor, 84
Colonial Garden Kitchens (catalog), 151
Comb, electric, 69
Combination golf ball/shoe cleat/iron groove cleaner, 99, 149
Combination padlock, 36
Combination screwdriver/pliers/hammer/triple wrench, 140

Computers, 129
Conference-call telephone device, 77, 146
Contact lenses, 65
Convertor-adaptor, 90, 147
Copiers and calculators, 80
Copper hoops, 12
Cordless electric lights, 38
Cordless Electric Tie Rack, 71, 145
Cordless extension telephone, 77–78, 146
Cork Pop, 54–55
Corklifter, 54, 143
Corkscrew, 54
Corn-kernel remover, 140
Counterfeit-bill detector, 80, 146
Couscous cooker, 46, 149
Croissant cutter, 141
Crossword puzzles, 97
Cue-ball combers with teeth, 128
Curler clamp, 69
Curling iron, 69

Da Vinci, Leonardo, 8–9
Damp Chaser, 6
Dashboard warning light, 84
Data-Timer, 81
Declaration of Independence, 9, 10
Defrost-Fast, 101, 150
Dent-U-Sonic denture cleaner, 63, 144
Dessert mold, 43
Dial-A-Balance, 80, 146
Dial-Down cigarette holder, 57–58, 144
Dial-O-Matic, 78, 146
Dial-Temp, 38, 140
Dial-X, 43, 142
Dialpack of contraceptive pills, 125
Dice, 97
Dione Lucas cooking show, 6

Disposables, 125
Divert-A-Matic, 77, 147
Do-it-yourself gadgets, 101–106
  where to find, 150–151
Doggie John, 104
Door chain, 36
Door jamb, 37
Double-broiler support, 47, 141
Dow Chemical Company, 62
Drain declogger, 40, 150
Drinking straws with a pump for milkshakes, 128
Drip coffee maker, 22
Drip-guard ice-cube trays, 53–54
Dripless Wine Pourer, 55, 143
Driver's warning system, 83, 148
Drying comb, 69
Duck press, 46, 141
Dumbwaiter, 9
Duran, Peter, 17
Dyna T-A-C, 79–80

Ecology, *see* Do-it-yourself gadgets
Edible tape, 128
Edmund Scientific Company (catalog), 151
Egg items, 48–49
Egg pick, 48, 141
Egg Thing, 48
Egg timer, 48
Egg wedgers, 49
Eggs-Ray, 48
Egypt (ancient), 7, 8, 18
Electric Airless Paint Gun, 41
Electric alarm mat, 37
Electric Barber, 91, 148
Electric Bingo Blower, 97
Electric brisker, 49, 141
Electric can opener, 4
Electric car, 148
Electric Cocktailmatic, 53
Electric Egg Cooker/Poacher, 48, 141

Electric egg inspector, 141
Electric eyes and timers, 35
Electric Food Smoker, 50, 141
Electric food warmer, 47, 141
Electric Foodmatic, 44, 141
Electric Fruit Juicer, 44, 141
Electric frypan, 19–20
Electric hair stylers, 91, 148
Electric heat pads, 81, 146
Electric heated towel stand, 67, 145
Electric Home Cheesery, 150
Electric hot-dog cooker, 49, 141
Electric ice-cream freezer, 150
Electric iron, 16
Electric Letter Opener, 74, 146
Electric page turner, 126
Electric paint remover, 41, 150
Electric Pants Presser Valet, 71
Electric paper shredders, 76, 146
Electric pencil sharpener, 75, 146
Electric Plate Warmer, 47
Electric poker machine, 97, 149
Electric potato peeler, 50, 141
Electric preset coffee starter, 141
Electric presser, 71, 145
Electric refrigerator defroster, 101, 150
Electric salad maker, 40, 141
Electric shaver, 68
Electric siren alarm, 37
Electric steak timer, 46, 141
Electric tape dispenser, 75, 146
Electric Teasmade, 50, 141
Electric Tie Rack, 71, 145
Electric Vegetable Juicer, 44, 142
Electro Sensor Panel, 85
Electrocardiogram while driving, 87
Electronic metal detector, 95–96, 149
Electronic mosquito chaser, 100, 149
Electronic thermometer, 65, 144
Emmett, Rowland, 89
Everything Tool, 140
Executive decision maker, 94, 149
Executive Jogger, 60, 144

Exerciser Exer-Gym, 144
Exerciser items, 60–63, 144
Extension mirror, 70
Eyebrow Comb, 5
Eyebrow Razor, 5

Facial sauna, 70
Family Dental Kit, 144
Figuretone exercise bike, 61, 144
Fish-cleaning board, 98, 149
Fish grill, 46, 142
Fish hooks with camera, 128
Fish poacher with removable racks, 46
Fish scaler, 98, 149
Fish skinner, 98, 149
Fishing net, 98, 149
Fishing thermometer, 98
Flavor Jector, 50, 142
Floramatic Plant Quencher, 90, 147
Flush toilet, 15
Folding cane, 91, 147
Folding ladder, 9
Folding shoe horn, 92
Folding wig stand, 92, 147
Food grinder, 18
Food mixers, 18
Foot grips, 38
Fork with crank, 128
Fountain pen, 73
Franklin, Benjamin, 9, 10–11
Freddie the Frog, 90
French egg poacher, 48
French-fry-potato cutter, 49
Fresh Vitamin C Fruit Juicer, 44, 141
Frigidaire, 53, 118
Fruit and lard presses, 13
Frypan cover, 47, 142
Fun, games, and sports, gadgets for, 93–100
    where to find, 149–150
Furnace alarm, 38, 139

Gadgets:
    for the bar, 52–56
    buying boom (in U.S.), 2
    collectors, 4
    development of, 1–22
    do-it-yourself, 101–106
    of fantasy and the future, 123–130
    first, 7
    for fun, games, and sports, 93–100
    for grooming, 66–70
    for your health, 57–64
    history of, 7–22
    for the home, 35–41
    for the kitchen, 42–51
    meaning of, 1–2
    for the office, 72–81
    publicity and promotion of, 6
    status role of, 3–4
    travel, 82–92
    where to find, 139–151
Garbage disposal, 105
Garlic press, 49
Gasoline stove, 13
General Electric Corporation, 64, 91
General Motors Corporation, 84, 85
General Precision Corporation, 68
Genie, 87, 148
George III, King, 95
Gillette Detangler, 70, 145
Gimbels (department store), 21
Gin-rummy card holder, 97
Glamor Tub, 15
Glass coffee maker, 22
GlassMaker, 56, 143
Goldberg, Rube, 96
Goldsmith Brothers (catalog), 151
Golf-ball heater, 99, 149
Golf practice range, 149
Golf score keeper, 99
Golf Swing Trainer/exercise club, 99, 149
Gourmet (magazine), 6

Gourmet Egg Cook, 48
Goury, André, 127
Grapefruit knives and spoons, 49
Gravy separator, 49
Greece (ancient), 8, 15, 16, 36
Grinder, 44
Grooming gadgets, 66–71
    where to find, 145

Hair detangler, 70, 145
Hair dryer, 69–70, 91, 148
Hair setters, 69
Hair Snare, 40
Hamilton, L. H., 20
Hamilton-Beach electric mixer, 20
Hammacher Schlemmer, 2, 3, 5–6, 49, 53, 123,
Hammacher Schlemmer (catalog), 151
Hammer with glass head, 128
Hand blender, 18
Hank's Pipe Cleaner, 59
Hard-boiled egg slicer, 49
Harmonica holder, 13
Hart, Harold, 151
Health Inhaler, 14
Health and medical items, 57–65
    where to find, 144
Hearing aid, 64
Heat diffuser, 47, 142
Hero Sheep Protector, 13–14
Hindsight coiffure mirror, 70, 145
His-Hers Rolls-Royce twosome, 87
Hitachi Corporation, 80
Hoffritz, Inc., 2, 5, 50,
Hoffritz, Inc. (catalog), 151
Hold-On, 147
Home Potato Peeler, 50, 141
Horn alarm, 37, 140
Hot Dogger, 49, 141
Hot roller, 69
Hot Weather Hat, 100, 149
House of Values (catalog), 151

Household items, 35–41
    categories of, 35
    where to find, 139–140
Hume, Jane, 16
Humidifiers, 106
Hunting for buried treasure, 95–96
Hush-a-Phone, 78, 146
Huxley, Aldous, 129
Hydraulic dental appliance, 63, 144

Ice chopper, 43
Ice-cream-cone holders, 128
Ice-cream freezer, 20, 150
Ice crusher, 18
Ice skates, with turn signals, 128
Immersion heater, 92, 147
Indoor toilet for pet, 104
Inkstands, 73
Inside-Outside window cleaner, 40–41, 139
Insomniac, aids for, 64–65
Insta-net, 98
Instant safety ladder, 38, 139
Interlingual Typewriter, 146
Intermatic Time-all Timer, 139
International Automobile Show, 88
International Business Machines Corporation (IBM), 78
International Patent Licensing Exposition, 60
Intruder door lock and alarm, 37, 140
Inventions, patent protection of, 10
Invento (catalog), 151
Invento alarm, 37, 140
Invento Cordless Triple Header, 74
Invento Prima Infra-Red Food Warmer, 47, 141
Isometric/isotonic exerciser, 61

Jay Norris (catalog), 151
Jefferson, Thomas, 9–10
Jogging board, 60, 144

Johnson, Nancy, 20
Johnson Smith Company (catalog), 73–74, 151
Juice-can opener, 50, 142
Juicer, 18, 44, 140, 142

Kerr, Graham, 6
Ketchup Saver, 49, 142
King Alfred Cutter, 12
King Roast Pin, 45, 142
Kitchen items, 5, 17, 42–51
 where to find, 140–143
Knauer, Virginia H., 102
Knife sharpeners, 43, 142
Kornbluth, Cyril, 89

Left Hand, The (catalog), 151
Left Hand, The (store), 4, 76
Left-handed tape measure, 75–76, 146
Lemon squeezer, 44
Lemon stripper, 5
Lighted medical spoon, 144
Lightning rod, 10
Lillian Vernon Company (catalog), 151
Liqui-Pour, 50, 142
Locators for fish, 98, 149
Locks and safety, *see* Household items
Lowrance Fish-Lo-K-Tor, 98, 149
Lucas, Dione, 6
Luggage wheels, 92, 147

Macy's (catalog), 151
Madame Shack's Dress Reform Abdominal and Hose Supporter, 14
Maglock, 36
Magna-Clean, 40–41, 139
Magnetic Windshield Protector, 86
Magnicator, 40
Make-your-own-fresh-creamer, 104, 150

Makeup mirror, 70
Mandolin (vegetable slicer/cutter), 6
Martinis, making, 53, 144
Masher, 18
Mazotas, Jack, 4, 6
Meat-chopper-and-salad-makers, 44
Meat thermometer, 45, 46
Meat injector, 50, 142
Mesopotamia (ancient), 7, 8
Message recorder, 77
Metal ball ricer, 43
Mexico (ancient), 7
Microwave oven, 102
Middle Ages, 8, 36
Miller, Louis, 95
Miner's candlestick, 12
Mini-Breaker circuit protector, 39
Mini-Cart, 81
Mini-Jet type drain unclogger, 40, 150
Miniature wrinkle remover, 147
Mini-Wonder, 37
Minimizer, 105
Mirrors, 70
Mitchell, Martha, 95
Mixers, 17, 20
Mobil Oil Corporation, 61
Mochica culture, 8
Money Monitor, 80, 146
Montgomery Ward's Catalog of 1883, 13
Mortar and pestle, 8
Motorola Corporation, 79
Muffin breaker, 49
Multilanguage typewriter, 72, 146
Multipurpose alarm, 37–38, 140
Murphy bed, 15
Musée des antiquités nationales, 8
Museum of Contemporary Crafts, 125
Musical desk calendar, 81
Mustache comb, 5, 65

Mustache cup, 69
Myrograph, 73

Napoleon I, 19
Nasal atomizer, 13
Naso-Vents, 65, 144
Natural-yogurt maker, 103–104, 151
Neapolitan coffee mill, 150
Nebel, Joseph, 125
Neck mirror, 70, 145
Neolithic period, 7
New York Times, The, 95
1984 (Orwell), 129–130
Nixon, Richard, 102
Noise-pollution meter, 105, 150
Norelco, 90
Nothing Box, 93, 149

Office of Alcohol Countermeasures (U.S. Department of Transportation), 84
Office items, 72–81
  where to find, 145–147
Oil and vinegar vials, 142
Olive spoons with handles, 43
Omlet lifters, 48, 142
Onion choppers, 49
Onion holders, 49
Onion and tomato slicer, 5
Orange peelers, 49
Orwell, George, 129–130
Oversize telephone, 95

Pacemaker, 63
Pack a clock, 89
Padded bottom on ketchup bottle, 128
Pair-O-Soks, 71, 145
Pants-presser, 90–91
Paper shredder, 76, 146
Papin, Denis, 19

Paris Inventors Fair of 1974, 126, 127
Parking-meter lamp, 95, 150
Parsley mincers, 49
Pasta makers, 104, 151
Pastry blender, 142
Pastry cutters, 43
Patek Phillipe's calendar watch, 126–127
PatExpo '74, 125
Patti-Stacker, 49
Peda-cycle exerciser, 61
Pencil gun, 74
  encil sharpeners, 18, 75, 146
Penholders, 73
Pepper mill, fish shape, 43–44, 142
Percolator, 21–22
Perc-O-Toaster, 18
Perfect Magic Arrow, 46
Personal breathing mask, 125–126
Peruvian Indians, 67
Phone Mate, 77, 146
Pickle fork, 43
Pineapple corers, 49, 142
Pipe smokers' ashtray, 58–59
Pizza in automobile, 128
Plug-tobacco cutter, 12
Pocket calculators, 80, 146
Pocket Grip Exerciser, 60
Pocket pepper mill, 92, 148
Pocket timers, 40
Pohl, Fred, 130
Poop-Scoop, 104
Pop-It Burger, 49, 142
Poplawski, Stephen J., 20
Portable Air Warmer, 100
Portable music stand, 9
Portabrella umbrella holder, 100
Posture developer, 61–62, 145
Potato baker, 47, 142
Potato-and-carrot serrators, 43
Potato curlers, 49
Potato masher, 43
Potter's wheel, 7
Poultry clips, 46, 142

Pratt Institute, 88
Pressure cooker, 17, 19
Princess Bust Developer, 14
Psychedelics, 94

Quick Change Dispenser, 86, 148
Quick-lime battery-run nail file, 71

Radar Sentry, 87
Radiant Heat Pad, 146
Radio Corporation of America (RCA), 39, 83–84
Radish roser, 49
Razor-blade sharpener, 68, 145
Reach-It, 47–48, 143
Rear-Window Defroster, 85
Refrigerator rotating shelves, 126
Remote-control garage-door opener, 87, 148
Remote electronic thermometer, 37–38, 140
Remote Mate, 77, 146
Remote telephone recorder attachment, 71, 146
Remote temperature check, 38, 140
Renaissance, 8–9
Rice cooker, 142
Roast grip holders, 45
Roast rack, adjustable, 45, 142
Rogers, D. A., 18
Rolling pin, three discs, 43, 142
Roman Empire, 8
Ronson Foodmatic, 44, 141
Rotary liquor dispenser, 143

Safety razor, 68
Safety Smoker Robot, 58
Salad driers, 46, 142
Salad makers, 18, 44, 141
Salad tossers, 46, 143
Salonette Beauty Mist, 70

Salt-and-pepper set, automatically clean, 44, 143
Salton Hottray, 47
Sausage stuffer, 13
Scalp stimulators, 70
Schick Company, 68
Schlumbohm, Dr. Peter, 22
Science and Mechanics, (magazine), 127
Scientific Martini Maker, 53
Scissor sharpener, 18
Scooper-scrapers, 49
Sears Roebuck catalog of 1885, 60
Sears Roebuck catalog of 1897, 14
Seat cane, 150
Self-coiling power cords, 38, 140
Self-correcting typewriter ribbon, 81, 146
Self-internal basters, 45
Self-stirring saucepan, 5, 143
Sellers, Peter, 95
Seltzer-bottle washer, 126
7-in-1 Combination-Pencil, 74
Shaving-cream dispenser, 68
Shoe remover, 126
Shoe stretcher, 102
Shoelace-tying machine, 126
Shopping cart with steering wheel, 128
Shower-All, 67
Shower-head jet spray, 66–67, 145
Shower Mate, 67
Silver, Leonard, 2, 3–4
Silver Eagle (racer), 88
Sip-Up toothbrush, 64
Skidometer, 97–98, 150
Skin-resistance, feedback, 65
Sleep sound devices, 64–65, 144
Sleeper (motion picture), 130
Slicers, 43, 45
Slim-A-Slice, 50
SmokeGard, 38, 140
Snail holders, 49
Snooper-Tronics, 95–96, 149
Snore stopper, 65, 144

Solid Oxygen System, 62, 144
Sonic Ear-Valves, 64
Sonus Switch, 94, 150
Spaetzle makers, 104, 150
Spaghetti fork, 49
Span-o-Matic razor-blade sharpener, 68, 145
Spenzal Tube Dispenser, 144
Sphygmostat Electric Blood-Pressure Minitor, 62, 144
Spil-Gard ice tray, 53, 143
Spin salad driers, 46
Spot-Shot, 98–99, 150
Spray stylers, 69
Spud Baker, 47
Stahly Razor, 68, 145
Standard Electric Kettle, 50
Staple remover, 75
Stapleless Paper Stapler, 75, 146
Steakwatcher, 46, 141
Steamette, 71
Steam/Press Valet, 71, 148
Stir and sip spoons, 49
Stone Age, 7
String dispenser, 12
Sugar-coated spoons, 128
Sugar dispensers, 49
Sumeria (ancient), 7
Sun-magnifying parasols, 128
Sunbeam Corporation, 20
Sunset House (catalog), 151
Sutton, Iris, 104
Swedish-meatball makers, 46
Swing Trainer, 99, 149
Swiss Army knife, 5, 140
Swiss drier, 46
Syphon pump, 40
Syria (ancient), 7
Syrup dispenser, 143

Tampone, Dominic, 3, 5, 53, 123–124
Tap water purification, 105
Tape dispensers, 75, 146, 147
Tape measure, 75, 147
Target launcher, 98, 150
Target practice light, 98–99, 150
Tater-Baker, 47, 142
Teen Tel Limiter, 147
Teflon, 51
Tele-A-Corder, 145–146
Tele-Caller, 77, 146
Telephone amplifiers, 78, 147
Telephone-call transfer service, 77, 147
Telephone-cap hold device, 147
Telephone Name Caller, 78–79, 147
Telephone Time Limiter, 77, 147
Television, 129
Tella-Cost, 102
Tenant's Emergency Indicator, 39
Tennis rebound net, 150
Tennis Teacher, 97
Thermo Pin, 45, 46
Thermo-Spoon, 49–50
Third arm, 47–48, 143
3-D films, 129
3M Remotecopier, 80
Throwaways, 125
Tiny Tim Keys, 128
TNS (pain control), 65
Toast warmer, 50, 143
Toaster, 17, 18–19
Toastmaster, 18
Tomato slicer, 49
Toothbrushes, 63, 64
Toothpaste dispensers, 63–64, 144
Touch and Cook electric ranger, 126
Touch-up Steam Press Valet, 71
Travel iron, 89, 90–91, 147
Travel items, 82–92
  where to find, 147–148
Tri-Fry pan, 51, 143
Tub Scrubber, 67
Turbo-jet for bath, 145
Turnpike toll gun, 86, 148

Twine holder, 12
Typewriters, 72–73

Ultra/Matic, 24, 80–81
Ultrasonic intruder alarm, 36–37, 140
United Society of Believers in Christ's Second Coming (Shakers), 10–11
U.S. Constitution, 10
U.S. Department of Agriculture, 19
U.S. Department of Transportation, 84
Universal Food Chopper, 44
University of Utah, 64

Vacuum cleaner, 16–17
Vegetable mincers, 49
Vegetable steamer, 5, 45
Venetian-blind cleaner, 39
Vibrating combs, 70
Video Voice, 150
Viking Snooper-Tronic, 149
Virginian stool shower, 15
Visible Writing Machine, 73

Waffle iron, 17, 19
Walker Universal Self-Puller, 54
Wallace, George, 65
Walter Drake & Sons (catalog), 151
Warm drinks, 54
Water filter, 105
Water massage, 66–67
Water-piks, 63
Water the plants, 90
Watergate affair, 76

Weight ladle, 50, 143
Westinghouse Corporation, 19–20, 67, 126
Whang, Kyu Bong, 60
Wheel, the, invention of, 7
Whimsymobile, 89
Whirlpool bath attachment, 67, 145
Whistling jar, 8
Wild oats sowing kit, 125
Windshield scraper, 85–86
Wine corks, adjustable, 55
Wine frigidaria, 55, 144
Wooden crepe spreader, 43
Wooden pepper mills, 42
Wooden skewer, 7
Wooden steak beater, 42
World of Henry Orient, The (motion picture), 94–95
World's smallest alarm clock, 91, 147
Worldwide (catalog), 151
Worldwide Electric Barber, 91, 148
World Wide Hair Stylist, 91
Worldwide immersion heater, 92, 148
Wrapping-paper dispenser, 12
Wrist Tape-Toter, 75, 147
Wristband Tote Tape, 75
Wristwatch phone, 79
Wrought iron napkin holders, 42

Xerox 400 Telecopier, 80

Yankee peddlers, 11
Yogurt makers, 103–104
Yolk separators, 49